Panayotis Papazoglou, PhD

An Educational Guide to the AVR Microcontroller Programming

First English edition (translation from Greek): 2018
Kessariani, Athens, Greece
email: **p.m.papazoglou@hotmail.com**

book photos: **Panayotis Papazoglou**

© Copyright 2018: **Panayotis Papazoglou**

ISBN: 978-1986008396

All rights reserved. No part of this book may be reproduced
in any form or by any means, without the author's written permission

Follow your dreams!

PREFACE

The type and the number of applications of the modern technology have driven to different solutions and approaches. Some of these solutions are based in a computer system which supports the corresponding requirements. The computers (PCs) that are used nowadays represent only one view of the reality, because they mainly support the requirements regarding the processing power, everyday applications, simplicity of using and programming, etc.

From the other point of view, there is a great number of applications that require an autonomous control system which is usually embedded in a device. In such systems, the requirements are focused on the operation autonomy, the physical size, the power consumption, etc.

Approaching these views from the aspect of architecture and programming, the core of the systems is a microprocessor or a microcontroller.

It must be noticed that the basic differences between microprocessors and microcontrollers, but also of their architecture and programming constitutes an unclear subject for many new engineering students and readers.

This book (volume 1) constitutes a complete basic educational guide which offers important knowledge and demystifies the AVR programming. Moreover, this book has been written by taking in account the real needs of students, teachers and others who want to develop AVR based applications.

All the programs and applications of the book have been developed and tested in a real microcontroller, in contrast with other books where the corresponding material has been developed only theoretically with no tests in practice.

The above lines, state the deep belief of the author that this book will constitute a useful teaching and educational tool for helping anyone understand the AVR applications. On the other hand, the book can be used by the teacher for organizing lectures and presentations as well as the laboratory exercises.

Panayotis Papazoglou, PhD

The book structure

The book (volume 1) consists of 7 chapters which include simple exercises or laboratory activities.

The **first chapter** is a general introduction to microcontrollers regarding their features and capabilities. The chapter also includes application examples that are focused on the corresponding components such as sensors, etc.

The AVR features such as the internal architecture, the memory system and the registers are presented in the **second chapter**. These features are presented only with the needed level of details in order to be easily understandable by the reader without any hard work.

In the **third chapter**, the basic instruction set is analyzed using a great number of figures. The assembly instruction set constitutes the basic tool for developing the corresponding applications. All the instructions are presented by using practical examples as well as special figures in order to demonstrate the corresponding operation. Thus, the reader understands the operation philosophy of every instruction.

Due to the fact that the algorithm constitutes the most important component for every application and the code synthesis is based on the structured programming concepts, a whole chapter (**fourth chapter**) has been written for presenting all this knowledge. Thus, the reader is guided step by step to the correct way of thinking for developing the desired application.

In the **fifth chapter** the reader learns for the first time how to exploit the digital input/output pins of the microcontroller. More precisely, the chapter begins with the basic knowledge on electrical circuits for understanding basic concepts as the current, voltage, etc. Next, the application of this knowledge in the external circuits of the microcontroller as well as the corresponding programming is analyzed.

The display units (e.g. seven segment displays, LCD) are analyzed in the **sixth chapter**. The display operation is very important for the user interaction with the application as well as for debugging.

Due to the fact that the modern applications support the input of complex data and instructions by the user, it is very important to present the corresponding switch circuits that can be used as keyboards. The **seventh chapter** analyzes how to develop such a circuit as well as the programming method for using it.

Important features of this book

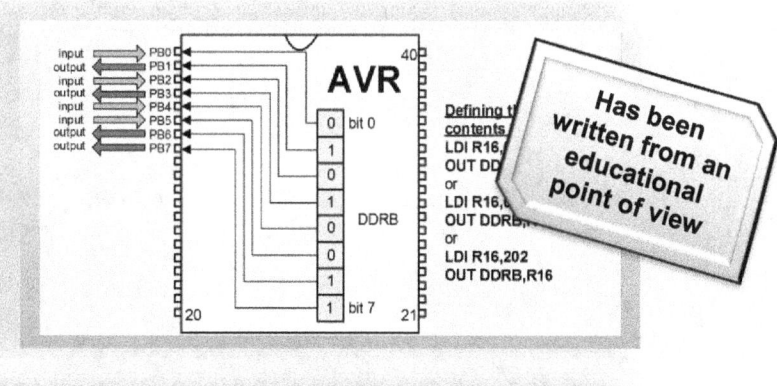

Has been written from an educational point of view

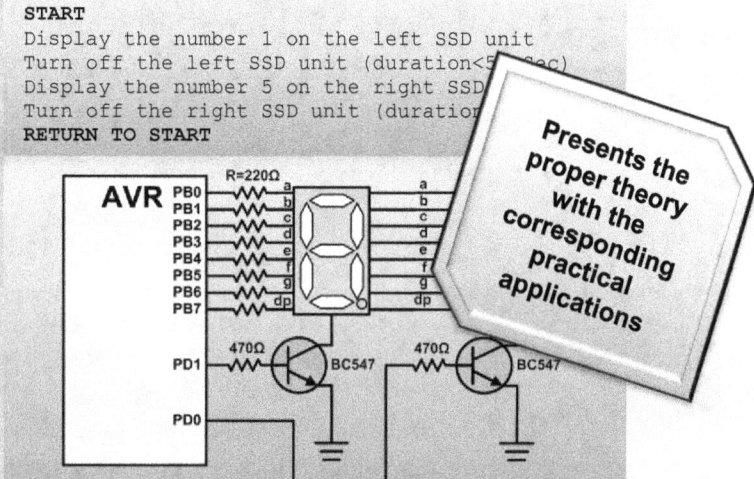

Presents the proper theory with the corresponding practical applications

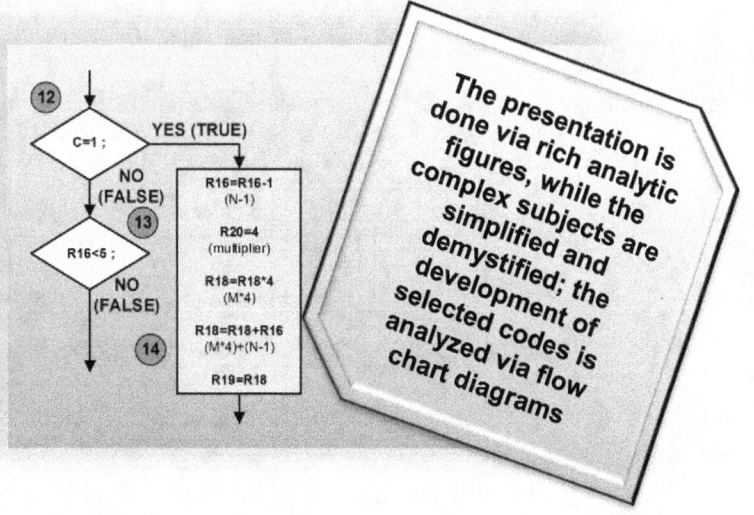

The presentation is done via rich analytic figures, while the complex subjects are simplified and demystified; the development of selected codes is analyzed via flow chart diagrams

```
;***************************
;Main program
;***************************
main:
  SBI PORTB,Button ;Pull Up r
                   ;activation
                   ;at pin PB0
  SBI DDRB,LED     ;The pin 1 of port B
                   ;will be set as output
scan:
        SBIC PINB,Button ;If the pin 0 (button)
                         ;is activated
                         ;(PB0=LOW=0V),
                         ;then the next
                         ;instruction
                         ;is bypassed

        RJMP scan        ;Return to check button
        SBI PORTB,LED    ;LED activation
```

Well documented source code

Step 2
Fill the following signal attributes based on the expected measurements on pin PD0.

T1 =

T2 =

T =

Frequency F =

Laboratory activities and simple exercises are included

Step 3
Connect one channel of the oscilloscope as described in the next figure. Set properly the time base (Time/Div) and the voltage scale (Volts/Div) on the oscilloscope in order to display the measured signal at the right size.

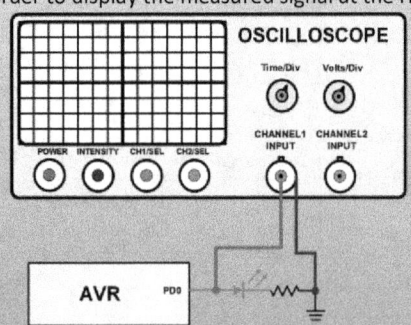

Important features of this book

Happy Reading!

Table of Contents

1. Introduction
1.1 General   1
1.2 Microprocessors and Microcontrollers   2
1.3 Microcontroller structural components   7
1.4 Microcontroller applications   9
1.5 Popular microcontrollers   13
Exercise Sheets   13

2. General Features of the AVR Microcontrollers
2.1 Introduction   17
2.2 Architecture   18
Exercise Sheets   27

3. The basic Assembly instruction set of the microcontroller
3.1 Introduction   35
3.2 Loading instructions   36
3.3 Arithmetic instructions   53
3.4 Logical instructions   65
3.5 More arithmetic and other instructions   69
3.6 Bit manipulation   77
3.7 Developing complete and functional programs   86
    3.7.1 Embedding INC files   86
    3.7.2 The basic Assembler directives   87
Exercise Sheets   89

4. Implementing basic programming structures
4.1 Introduction   101
4.2 Comparison and branch instructions   101
    4.2.1 Basic programming structures   101
    4.2.2 Implementing single control flow structures (selection)   103
4.3 Iteration structures   112
    4.3.1 Iteration structure do-while   112
    4.3.2 Iteration structure while-do   114
    4.3.3 Nested loop (iteration structure)   115
    4.3.4 Implementing a loop without comparison   118
    4.3.5 Defining the iterations number beyond the register limit   119
4.4 Absolute and relative jump   121
4.5 The stack   124
4.6 Time features of code execution and delay programs   128
4.7 Macroinstructions   132
4.8 Program coding in memory   135
Exercise Sheets   137

5. Basic Programming of the Input/Output (I/O) Ports
5.1 Introduction   **145**
5.2 Elementary electrical-electronic circuits   **145**
    5.2.1 Circuit with resistor   **145**
    5.2.2 Circuit with diode   **147**
    5.2.3 The LED diode   **148**
5.3 Basic manipulation of digital ports   **150**
5.4 LED manipulation through microcontroller ports   **157**
5.5 General purpose switch circuits   **158**
    5.5.1 Operation methods   **158**
    5.5.2 Simplified switch circuit   **159**
5.6 Implementing a time delay   **159**
    Application 5.1 – Activating eight LEDs   **161**
    Application 5.2 – 8 LED activation and deactivation with time delay   **161**
    Application 5.3 – Successive LED activation with logical shift   **163**
    Application 5.4 – Traffic light control   **165**
    Application 5.5 – Controlling a LED with a button   **169**
    Application 5.6 – Button circuit simplification
    using the internal Pull-Up resistor   **172**
    Application 5.7 – Direct LED control using buttons   **173**
LABORATORY EXERCISE 1 - Basic electrical circuits   **176**
LABORATORY EXERCISE 2 - Time Delay   **182**
LABORATORY EXERCISE 3 - Switch circuits   **184**

6. Display units manipulation
6.1 Seven segment display manipulation   **187**
    Application 6.1 – Testing the SSD unit operation   **189**
    Application 6.2 – Displaying the numbers 0 to 9   **192**
    Application 6.3 – Controlling the SSD unit
    by using multiplexing   **194**
    Application 6.4 – Display four digits on multiple SSD units   **197**
    Application 6.5 – Automated three digit number display   **201**
6.2 LCD screen   **214**
    Application 6.6 – Using the LCD screen (16x2)   **217**
LABORATORY EXERCISE 4 - Seven Segment Display (SSD) units   **228**
LABORATORY EXERCISE 5 - Using an LCD 16x2 screen (1602)   **231**

7. Switch circuits for user input
7.1 Introduction   **233**
7.2 The keyboard layout   **233**
    Application 7.1 – Developing a keyboard for displaying
    the digits 0 to 7 on SSD units   **236**
    Application 7.2 – Displaying the digits 0 to F on SSD units
    using the keyboard   **247**
LABORATORY EXERCISE 6 - Keyboard development   **255**

Appendix A - Simulating assembly source code in Atmel Studio 7 **257**
Appendix B – Instruction summary **265**

1 Introduction

Content-Goals
This chapter presents the operation of microcontrollers in general as well as some corresponding indicative applications. The above operation is analyzed in comparison to the classical microprocessor and the conventional computer as a system. The goal is to clarify the microcontroller's role in the application development process of specific purpose.

Chapter contents
1.1 General
1.2 Microprocessors and Microcontrollers
1.3 Microcontroller structural components
1.4 Microcontroller applications
1.5 Popular microcontrollers

1.1 General

Microcontrollers are usually embedded in autonomous systems which support specific applications. An electronic thermometer, a robotic vehicle, a smart television, the electronic controller of a car, a mobile medical instrument, the bus ticket controller, the metro security system, the automated transaction systems, the traffic lights controller and many more, constitute systems where an embedded microcontroller supports the corresponding service. It is unlikely in such a system for a traditional PC to be used. This happens due to hard restrictions such as the physical size, the operation autonomy, the energy consumption, the difficult installation conditions, the real time response to external events, the reboot time, etc. The above constraints cannot be met by a traditional PC.

Thus, the microcontroller as a single electronic chip is embedded in such a system in order to satisfy the above constraints. A microcontroller is usually called "small computer" due to the fact that it can autonomously support the above applications.

On the other hand, humans are more familiar with traditional PCs, and thus the corresponding introduction will start from them. Using initially the microprocessor as a reference point, the presentation will proceed to the microcontroller in order to be able to make a comparison. With this approach, many of the basic microcontroller features (which constitute the core of the book) will be more understandable.

1.2 Microprocessors and Microcontrollers

A computer system (Computer-Personal Computer-PC) consists of various components (internal and external) in order to be exploited by the users that select the corresponding applications. A computer does not only contain a microprocessor and a memory system but also has output units such as a monitor to display information, input units such as a keyboard for entering data, etc. Moreover, the computer consists of supplementary circuits for implementing the communication between the main components (e.g. microprocessor, memory) in order to support the instruction execution.

Figure 1.1 shows a typical structure of a computer. The computer operation as a system is based on the interaction between the three main components (internal units such as microprocessor, memory and external units such as monitor, keyboard). The above components constitute independent units which are interconnected through the bus system.

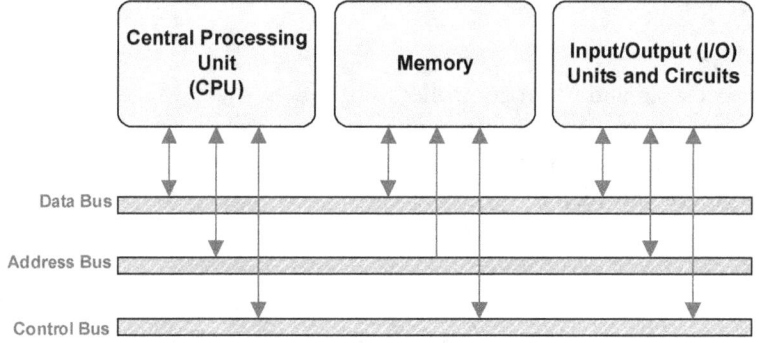

Figure 1.1 A typical computer structure

More precisely, the goal of each component (unit) of figure 1.1, is:

Central Processing Unit -CPU. The CPU accepts instructions from the main memory and implements the execution process for producing the corresponding results. If for example the programmer has used an instruction for addition (code in memory), then this instruction is transferred in the CPU for performing the addition (instruction execution) and the result is returned back to memory. The CPU is implemented in an integrated circuit (IC-chip) which is called **microprocessor**.

Memory. The memory (main memory) stores the programs and the corresponding data that are processed by the programs. The memory constitutes an one dimensional array with identical locations in terms of capacity. On the other hand, every memory location has a unique identifier which is called address. For reading or writing to a memory location the corresponding address has to be activated.

I/O units. The I/O units ensure the system functionality and the communication with the outer environment (outside world) in order for the computer to be useful to the user. A well-known example is the keyboard for entering data to the computer.

Data bus. The system bus is common among the computer components in order to control and exchange data between the system units. The data are transferred via the data bus. For example, the result of the mentioned addition is transferred to the memory via the data bus.

Address bus. Addresses are unique numbers that are assigned to all computer components for supporting the corresponding communication. The result of the previous addition will be used again as an example. The result storage location in the main memory is activated through an address which is transferred via the address bus.

Control bus. The data exchange between computer components is achieved via special control signals that are transferred through the control bus. These signals ensure that a data packet will be delivered at the correct address (destination) and the corresponding units will be activated in the right priority.

From all the above it is obvious that the microprocessor is an integrated digital circuit which only contains processing elements for executing instructions from the memory. That means that the microprocessor cannot operate autonomously. In other words, the exploitation of the microprocessor cannot be done without the support of additional components such as the memory, I/O units, etc. Figure 1.2 shows from a different point of view the independent components (units) of the computer beginning from the microprocessor level towards to the outer environment (outside world). Figure 1.2 shows also that the exploitation of the microprocessor is a complicated process due to the fact that much more components have to be combined in order to form a complete computer system. The above constraints increase the system cost, the management complexity and the installation difficulties for an autonomous application.

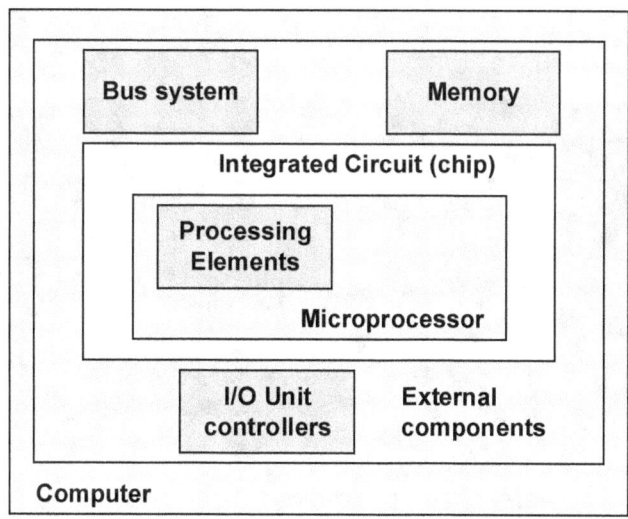

Figure 1.2 The microprocessor and the system components

Figures 1.3a and 1.3b, show the front and rear views of a real microprocessor IC respectively.

Figure 1.3a Front view of a microprocessor

Figure 1.3b Rear view of a microprocessor

A microprocessor offers high processing performance by executing instructions in short times. The fact that the microprocessor only executes instructions means that there are many supplementary components and circuits needed to build a functional system.

Figure 1.3c shows a main board (motherboard) which hosts the microprocessor as well as the supplementary components and circuits for the system operation.

Figure 1.3c Main board (motherboard) for hosting a microprocessor and other supplementary units and circuits

On the other hand, the integrated circuit of a microcontroller (fig. 1.4) contains internally all the components that in the case of the microprocessor were built externally (e.g. memory, I/O controllers).

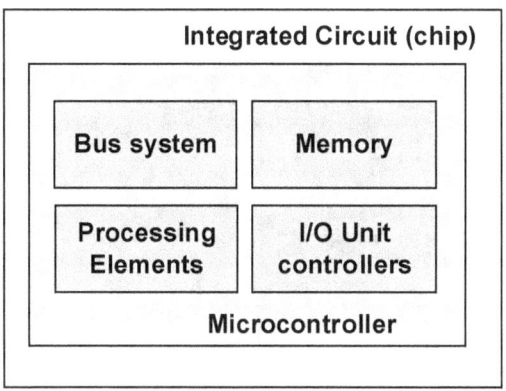

Figure 1.4 General microcontroller structure

The existence of the needed components inside the microcontroller, makes possible the autonomous operation of the corresponding system by combining the low cost, the limited complexity as well as the installation simplicity in the final applications.

Thus, a microcontroller is implemented in a single integrated circuit with some dozens of pins. Figures 1.5a and 1.5b show the microcontroller models ATmega328 and ATmega32 respectively.

Due to the fact that the microcontroller contains the needed components for autonomous operation (fig. 1.6), the corresponding external circuits (which are much simpler than the supplementary units and circuits of a microprocessor) consist only of electronic components for the power supply, the reset circuit and the optional clock oscillator (the microcontroller can be operated only with an internal clock in a lower frequency).

Figure 1.5a
The microcontroller ATmega328

Figure 1.5b
The microcontroller ATmega32

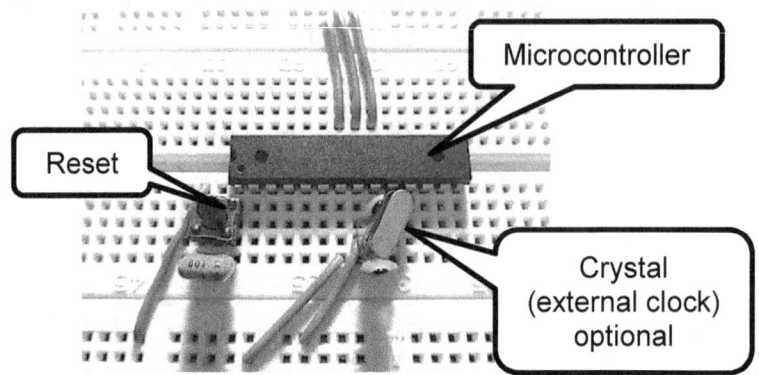

Figure 1.6 External microcontroller circuit

This simplicity makes the microcontroller an ideal solution for mobile and other applications where it can be embedded to support autonomous operation by implementing the control tasks of the target application.

From the previous brief presentation it is obvious that the microprocessors and the microcontrollers start from different approaches and are developed for applications and systems of different purpose.

Table 1.1 summarizes the differences between microprocessors and microcontrollers.

Table 1.1 Basic differences of microprocessors and microcontrollers

Microprocessor	Microcontroller
An integrated circuit which only implements the central processing unit (CPU) and does not constitute a computer	Constitutes a "small computer" because it contains processing unit, memory, I/O controllers, etc.
Is used in general purpose systems that support many and complicated applications	Is used in systems where control is needed as well as small application construction size, low energy consumption, low cost, fast response in external events, etc.
Is used in cases where a high processing performance is needed. Offers high performance where complicated tasks and calculations are needed	Performs simple tasks and operations, while the processing performance is much lower as compared to microprocessors
Constitutes the heart of the modern computers (PCs)	Is ideal for the embedded systems (constitutes the heart of the autonomous systems like the everyday appliances)
Supports complicated mathematical calculations (e.g. contains a floating point unit)	Does not support such capabilities. Some calculations are only supported by special software
It has high cost (is also based on many complementary units and circuits to form a computer system)	The cost is extremely low (starting from $2-$3)
High power consumption (also due to the complementary units and circuits)	Very low power consumption (can be supplied from a single battery)
High clock frequency (e.g. 1-3GHz)	Low clock frequency (e.g. 1-30 MHz)

Higher delay to external event processing due to multiple software levels existence (e.g. operating system)	Has been designed especially for fast response to external events. Is ideal for processing sensor and other device data in real time
Does not used in real time systems due to software delay issues	Is ideal for real time systems where the results and the decisions have to be produced in limited time slices
The final system has a large physical size and thus cannot be used in size limited applications	The final system has a small size and thus constitutes an ideal solution where the available physical size of installation is limited
Long booting time	Very short booting time

Note
The previous table refers to typical cases of microprocessors and microcontrollers. There are cases of advanced microcontrollers with higher cost which offer higher processing performance as compared to some microprocessors.

1.3 Microcontroller structural components

Due to the fact that a microcontroller supports a great range of autonomous applications, it is obvious that it has the capability of communicating with the external environment ("outside" world) for exchanging and processing the corresponding signals in order to satisfy the applications requirements. For efficiently supporting the above capabilities, the microcontroller contains a number of components (units) such as Control Unit (MCU), memory, interrupt controller, signal converter, etc. As mentioned before, microprocessors are used in general purpose systems, while microcontrollers are used in special purpose systems. Figure 1.7 shows with more detail the components (units) that constitute a typical microcontroller.

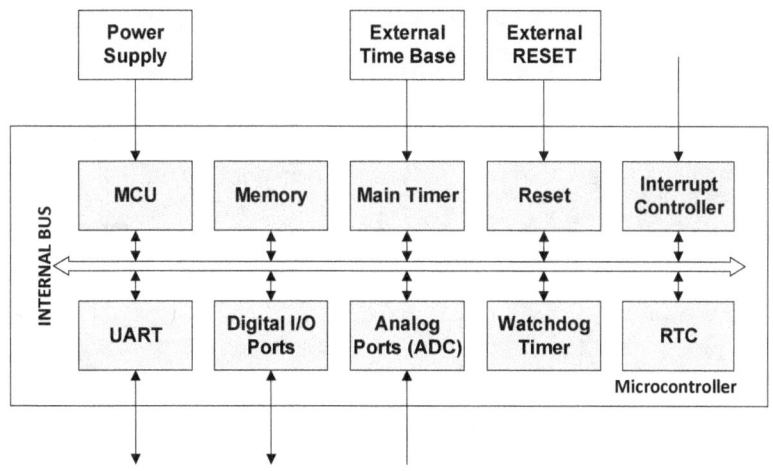

Figure 1.7 Basic microcontroller units

The basic operation of the microcontroller units in figure 1.7, is:

- **MCU (Microcontroller Control Unit)**. The MCU executes the program instructions that are stored in the program memory. Based on the internal architecture, there is a corresponding Assembly instruction set. The MCU operation requires registers (general and special purpose), arithmetic and logic unit (ALU), etc, like a typical computer CPU.

- **Memory**. The memory stores programs and data and can also be connected with an external memory unit. There are many types of memories that are used in microcontrollers such as:
  - EEPROM (reprogrammable with electrical signals)
  - Flash (fast storage)
  - Classic RAM (random access)
- **Main Timer**. The instructions execution as well as the internal units interaction for supporting a functional microcontroller is implemented using a common clock signal which is generated by a timer (oscillator). This oscillator can be internal or external (using a crystal for higher frequencies).
- **Reset**. Operation restart via an external circuit or internal program
- **Interrupt controller**. This controller implements the external event manipulation by interrupting the main program execution flow for a tiny time slice in order to execute the corresponding interrupt service routine which supports the event. If for example a switch for lighting a LED is momentary activated while the main program is under execution, then the interrupt controller activates the corresponding code section in order to light the LED and after a tiny time slice the execution flow returns to the main program.
- **UART (Universal Asynchronous Receiver Transmitter)**. This unit supports the asynchronous serial communication which is used for controlling external devices as well as data exchanging between the microcontroller and other devices. The serial communication includes the data preparation and framing for ensuring the correct reception at the destination.
- **Digital I/O Ports**. Every microcontroller has a set of ports that can be used as inputs or outputs. The signals that can be recognized by these ports are exclusively digital, that is signals of a 0 or 5V level (fig. 1.8).

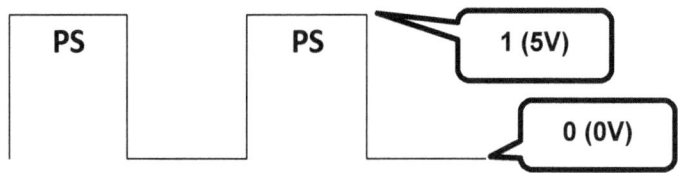

Figure 1.8 Digital Signal (PS=Positive Segment=5V)

- o **PWM**. Any device that can be activated or deactivated with a digital signal, can also be controlled by a microcontroller. Another feature of the microcontrollers is the pseudoanalog signal generation for controlling analog devices. These signals are generated using the PWM (Pulse Width Modulation) capability of the microcontroller which is based on controlling the duration of the PS (Positive Segment) in relation with the signal period (Duty Cycle-DC). Figure 1.9 shows some waveforms with different DC that are generated by the digital outputs. By setting the DC percentage in combination with a high signal frequency (e.g. 500Hz), a different mean output voltage value can be achieved. Thus, intermediate voltage levels in the range [0,5]V can be produced. For example, a motor's control can be achieved using the above technique.

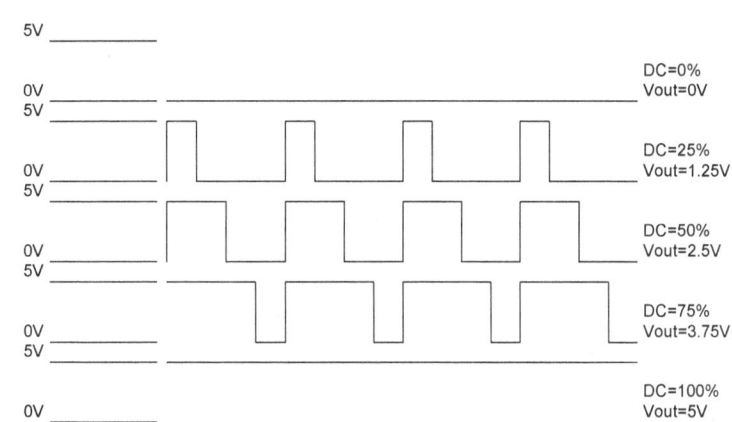

Figure 1.9 Some PWM indicative signals

- **Analog Ports (ADC)**. Many microcontrollers contain special converters (Analog to Digital Converter-ADC) which convert analog signals to digital. These ports usually read analog signals in the range [0,5]V, while this can be extended using a voltage divider. The resulted digital signal is represented in bits where the corresponding number of bits defines the conversion accuracy or resolution. If for example, the microcontroller uses 10bits resolution, then the resolution step is 5V/1024=0.0048V (2^{10}=1024).

- **Watchdog Timer**. Due to the fact that the systems that are based on a microcontroller have been designed to operate autonomously in the physical field of installation there must be a way the application operation to be recovered from a severe error (deadlock). For achieving this, a special counter which is reset periodically (through the application if there is no problem) is activated. Otherwise, if the counter exceeds a predefined limit (this means that the counter has not been initialized by the normal operation of the application) then it implements a reset to the microcontroller to recover the normal operation.

- **RTC (Real Time Clock)**. In some microcontroller models, special circuits are used for keeping the real time. Despite the existence of such circuits, the corresponding time is not accurate. Thus, external circuits or devices are used.

1.4 Microcontroller applications

As mentioned before, a microcontroller operates autonomously like a "small computer". The development of a full functional application which is fully controlled by a microcontroller, requires the combination of the most suitable external circuits, sensors, etc. Figure 1.10 shows a set of external "components" that can be combined in order to build the desired application.

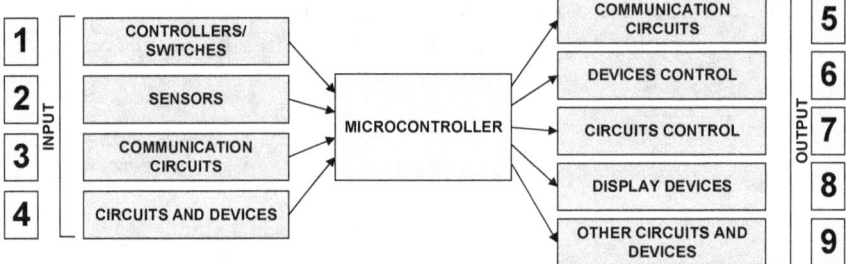

Figure 1.10 Component types for application development

Application example 1: Temperature display

There are already known features of a microcontroller and thus, this example is focused on the features and development of the application. This application measures the room temperature which is displayed through LEDs (the LEDs correspond to different temperature levels). The application of the first example consists of the following components:

(a) Temperature sensor (connected to an analog input of the microcontroller)- component type 2 (fig. 1.10)

(b) 3 LEDs (connected to digital outputs of the microcontroller through resistors) –component type 8 or 9 (fig. 1.10)

(c) A program which reads the sensor and controls the LEDs

Figure 1.11 shows the application circuit. For simplicity reasons, it is assumed that the three LEDs are lit according to the temperature ranges [10,19.99], [20,29.99] and [30,39.99] °C degrees respectively.

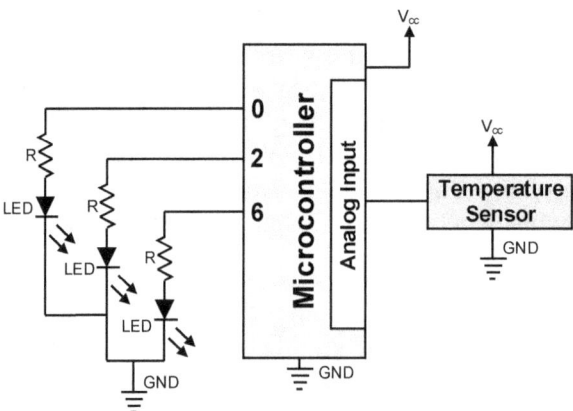

Figure 1.11 Application for displaying temperature

The following algorithm describes the logic of the corresponding program for displaying the temperature.

REPEAT
```
Turn off all the LEDs (write to digital outputs 0b00000000)
```

```
Read temperature sensor (T) from the analog port
If (T) belongs to [10,19.99]
    then write to digital outputs 0b00000001
Otherwise
    If (T) belongs to [20,29.99]
        then write to digital outputs 0b00000100
    Otherwise
    If (T) belongs to [30,39.99]
        then write to digital outputs 0b01000000
WHILE There is power supply
```

Of course, a scale conversion has to be done from the range [0,1023] (which is the output of the ADC with 10bits) to the corresponding levels which will be checked by the program. The symbol 0b means that the number is binary, where each bit represents the state of a pin in the output port of the microcontroller. Thus, if the outputs are 8, numbered from 7 to 0, then the sequence 0b00000001 means that the pin 0 where the first LED is connected will be set to 5V (active for the temperature range [10,19.99]). Due to the 5V voltage, a current will flow in the LED circuit and the LED will be lit.

In a little different application, the temperature can be displayed in seven segment display units (fig. 1.12). The microcontroller reads the analog output of the sensor and converts the signal to digital form. The digital data (temperature measurement) are converted to discrete digits which are sent to the seven segment display units in real time.

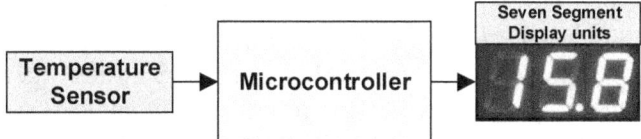

Figure 1.12 Application for displaying temperature using seven segment display units

Application example 2: Controlling the air-condition

An application for controlling the air-condition remotely (wireless communication) can be developed (fig. 1.13) by combining the proper components. The corresponding settings are defined by the user. The current temperature (measured by the application) is compared with the user temperature settings. Based on this comparison, the air-condition is adapted to the needed operation requirements. Thus, the room temperature remains stable or is adapted to the new temperature level which is defined by the user.

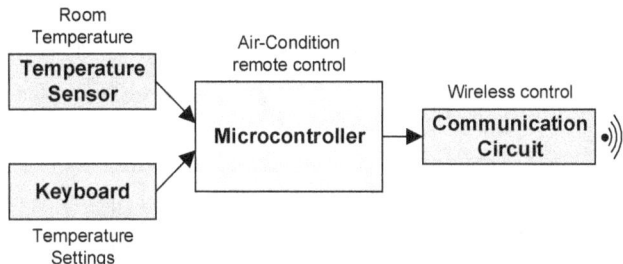

Figure 1.13 Controlling an air-condition

Application example 3: Controlling a robot

Due to autonomous operation with simple circuits and the small size, the microcontrollers, are ideal to be embedded in mobile systems such as an autonomous robot. In this example, the "control center" of the robot controls the motor's activation according to signals that are sent by the light and distance sensors (fig. 1.14). The microcontroller represents the "control center" and is programmed to make a combined processing of the sensors measurements and the remote control instructions in order to decide for the robot movement. The movement is controlled by small signals that are sent from the microcontroller to the motor control circuits.

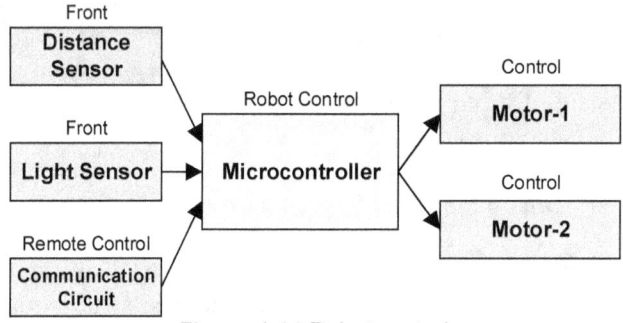

Figure 1.14 Robot control

Application example 4: Car controller/computer

Modern cars display illuminated indications to the driver about the car status (fig. 1.15). The above status is related to special adapted sensors that are installed in critical points of interest in the car. Exploiting the low cost of the microcontrollers and based on the requirements for updating the status indications in real time, a different microcontroller can be used for every sensor group. Thus, the processed measurements are sent to the central microcontroller which constitutes the core unit of the car controller/computer in order to display indications and to store faults for later processing by the service personnel. Of course, a different architecture with only one microcontroller and many sensors can be implemented.

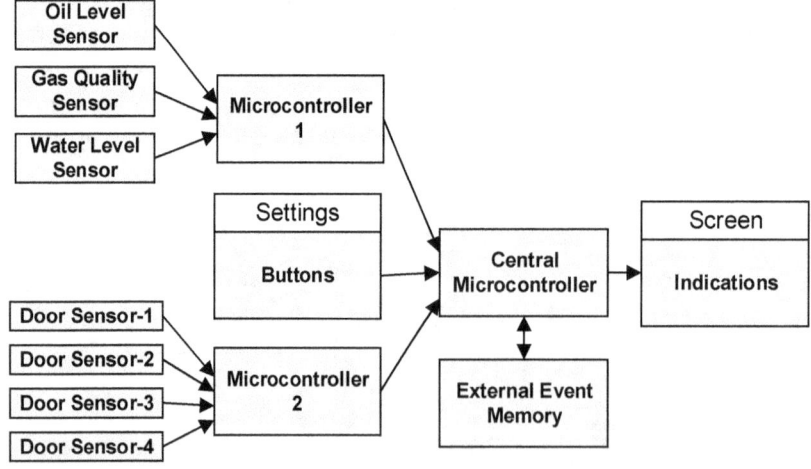

Figure 1.15 Car controller/Computer

1.5 Popular Microcontrollers

From all the above, it is obvious that there is a huge number of microcontroller applications. Due to everyday applications and other advanced services, many companies offer microcontrollers that are available to engineers and hobbyists. An engineer has to choose the most suitable microcontroller for developing the desired application.

Table 1.2 shows the most popular microcontroller models.

Table 1.2 Popular microcontrollers (indicative)

Company	Internet url	Indicative model
Texas Instruments	http://www.ti.com/	MSP430F1x
Microchip	http://www.microchip.com/	PIC16F84A
Zilog	http://www.zilog.com/	S3F80P5
Freescale	http://www.freescale.com/	HC08EY
Atmel (now Microchip)	http://www.atmel.com/	AVR ATmega328
Parallax	http://www.parallax.com/	Basic Stamp
Maxim	http://www.maximintegrated.com/	MAXQ611

The most popular models are offered by Microchip and Parallax.

EXERCISE SHEETS
Knowing the microcontrollers

1. Note every phrase as right or wrong

Phrase	Right	Wrong
The microcontrollers are usually embedded in autonomous systems	☐	☐
The processing unit in microcontrollers and microprocessors executes the program instructions	☐	☐
The memory is a two dimensional array for storing addresses and data	☐	☐
The microprocessor operation requires supplementary external circuits	☐	☐
The board for supporting a microcontroller contains more complex circuits than the board for supporting a microprocessor	☐	☐
The autonomous operation of the microcontroller requires an extremely simple external circuit	☐	☐
Microcontrollers have in general higher cost than the microprocessors	☐	☐

14 ☐ CHAPTER 1

Microprocessors are ideal for embedded systems	☐	☐
Microprocessors have in general lower power consumption than microcontrollers	☐	☐
The digital I/O ports of the microcontroller recognize signals in the range [0-10]V	☐	☐
The PWM technique is used for controlling analog devices	☐	☐
More bits in the ADC unit means higher resolution	☐	☐

2. Search in the literature the frequency and period definition for a periodical signal, as well as the corresponding prefixes for expressing the numbers' magnitudes easier. Based on the above, fill the following tables.

Hz	KHz	MHz	GHz
		1	
16300			
			1.3
	345.8		
		24	
1			
	1		
			1
		4.77	
	20.8		

nSec	μSec	mSec	Sec
			1
	1		
		23.1	
1300			
			5
		4	
	250		
43789			
	1000		
		1000	

Frequency	Period
15KHz	
	1μSec
16MHz	
	0.3μSec
1GHz	
	1Sec

1Hz	
	1mSec

3. The Duty Cycle is the duration of the positive segment (PS) as compared to the signal period (is expressed as a ratio). If the logical 0 is 0V and the logical 1 is 5V, fill the following tables.

Duty Cycle	Pseudoanalog voltage
32.2%	
	4.45V
50%	
25%	
	4.75V
11%	
	3.75V
100%	
	5V
	0V
0%	
	1.25V
95%	
	2.25V
22.5%	

4. Draw in the scaled paper the waveforms for the following Duty Cycles (DCs) and calculate the corresponding mean voltage V_{out}.

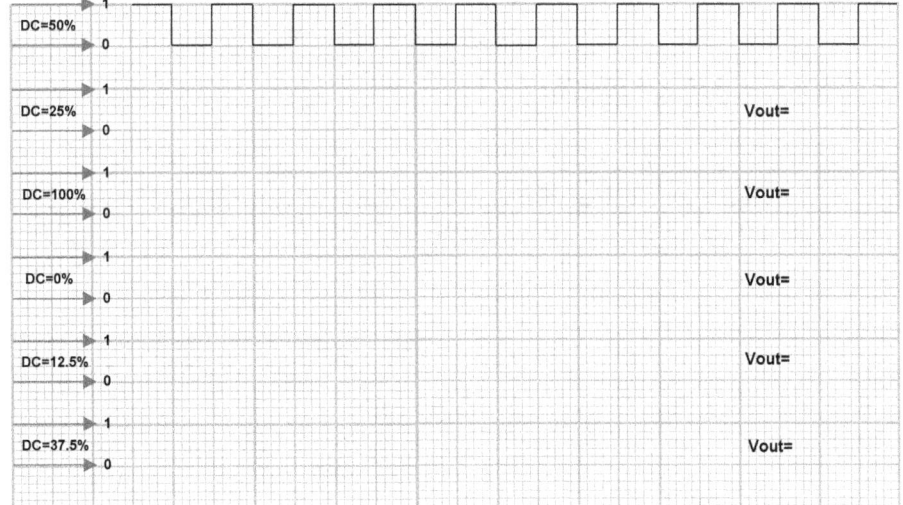

5. Combine the components of figure 1.10 and fill the following table.

Input components	Application	Output components
	Bus ticket control system	
	Microwave oven	
	Television remote control	
	Automated bank transaction machine	
	Autonomous robotic vehicle	
	Mobile phone	
	Digital distance counter for bike	

Download material

http://panospapazoglou.gr/support/

Email

p.m.papazoglou@hotmail.com

2 General features of the AVR microcontrollers

Content-Goals
This chapter presents the architecture and the features of the AVR microcontrollers. The above information is absolutely necessary in order to exploit the microcontroller through the corresponding programming. The architecture that will be presented is the same for all the classic 8bit AVR models.

Chapter contents
2.1 Introduction
2.2 Architecture

2.1 Introduction

There is no doubt that the Atmel AVR microcontroller family is the most popular in the world community. This book refers to the 8bit AVR models of the family which are used widely in many platforms such as the Arduino and other applications. As will be shown, an 8bit architecture is satisfying for the corresponding applications. Recently, the Atmel released the 32bit microcontroller family. Moreover, there are many available microcontroller platforms in the market that can be used for developing rapidly applications by anyone.

Regarding the microcontroller autonomy, the following operations have to be supported:

(a) Instruction execution from the program memory (through MCU)
The MCU contains various units such as Arithmetic and Logic Unit (ALU), general and special purpose registers, decoder, etc., for supporting the program instruction execution of the corresponding application.

(b) Exchanging signals with the outer environment (through the ports/pins)
The digital and analog ports support the communication with the outer environment. Special significance have the digital ports for controlling an external device. On the other hand, every analog input recognizes signals in the range [0-5]V and converts them within the program in 1024 (10 bits) different levels. With a voltage reference (AREF) of 5V, a resolution 5/1024=0.0048V is supported, while can be adapted to current needs using a different reference voltage for higher accuracy.

(c) Autonomous operation (power supply and timing)
The autonomous operation does not refer only to the external power supply but also to the timing which ensures the instructions execution as well as the normal

microcontroller operation. The timing is referring to the main clock of the microcontroller as well as the supervision operation that has been presented in the previous chapter. In most cases, the timing is based on an external clock (crystal) with a typical frequency of 16MHz.

Figure 2.1 shows the general organization of a typical microcontroller as mentioned in the previous chapter. The most important units are related to instructions execution for supporting a functional microcontroller. Programs and data are stored in the memory system, while the instructions are executed by the MCU.

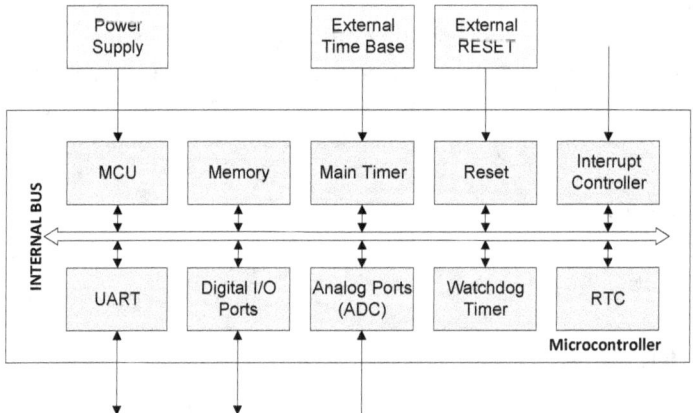

Figure 2.1 Microcontroller structural components

2.2 Architecture

Figure 2.2 shows the internal architecture of the microcontroller. The architectural components that are related to the MCU such as Arithmetic and Logic Unit (ALU), registers, program counter, etc., do not have significant differences as compared to the corresponding components of a microprocessor. Moreover, the microcontroller consists of additional units (except MCU) for supporting other operations such as the communication with external devices, the supervision, the analog signal measurement, etc.

Figure 2.2 is focused on the internal architecture of the microcontroller where the main components are related to the MCU.

The AVR microcontroller is based on the Harvard architecture which means that uses different memory and bus for programs and data respectively. This feature offers the capability to prepare the next instruction (recalling from memory), while the current instruction is under execution. Thus, the instruction execution is achieved in one machine cycle (MC). It has also 32 general purpose registers which can be used more flexibly in order to manipulate more efficiently the data memory. Program memory, is a flash memory which ensures the fast reprogramming while is divided in two areas, the boot area and the program area.

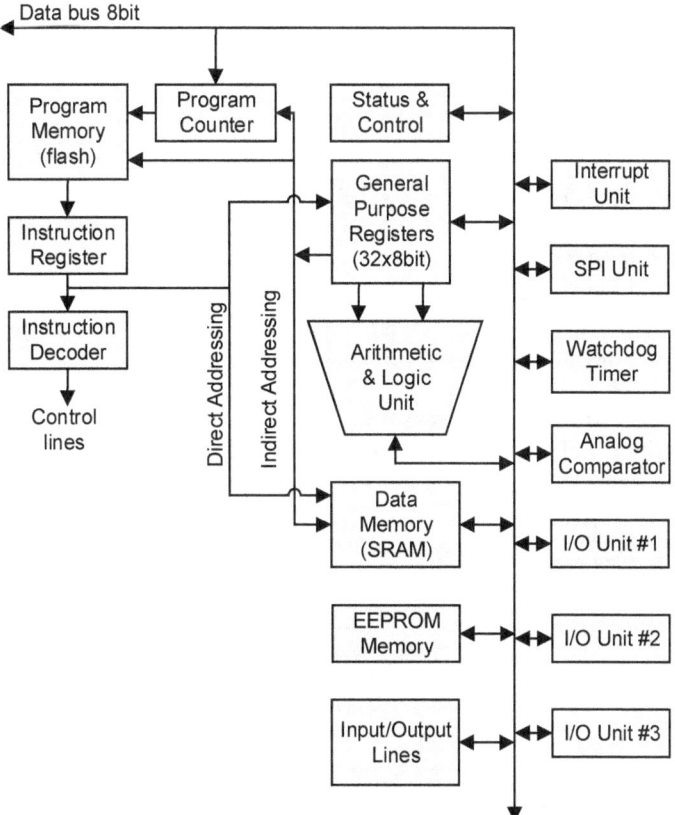

Figure 2.2 Internal architecture of the microcontroller

The components that are shown in figure 2.2 are generally analyzed as follows:
- **Flash Memory.** Hosts the user programs, while the content of the memory remains stored during power off period. This memory is reliable for approximately 100.000 program uploads.
- **Program Counter.** Points to the address of the next instruction that is going to be executed by the MCU.
- **Instruction Register and Decoder.** Are storing and Analyzing the form and the operation of the current instruction in order the needed signals to be generated ensuring the corresponding execution.
- **Status and Control.** Indicate the ALU status regarding the produced result.
- **General Purpose Register (32 registers).** These registers are available to the user in order to develop functional programs including flow control, arithmetic operations, etc.
- **Arithmetic and Logic Unit (ALU).** Execution of arithmetic and logical operations.
- **Data memory.** Storing and reading data that are manipulated by programs.
- **EEPROM Memory.** Permanent data storage with erasing capability.

- **Interrupt Unit.** A special circuit which recognizes the external interrupts (events) and helps the corresponding manipulation by the MCU.
- **SPI Unit.** Supports a special communication protocol for exchanging data with external devices in digital form.
- **Watchdog Timer.** The main timer ensures the internal units cooperation in order to support the instructions execution. The watchdog timer operates supplementary in order to support the microcontroller normal operation after a malfunction situation.
- **Analog comparator.** Supports the operation of reading analog signals.
- **Port sub-Units (I/O units).** Support the data exchange between the microcontroller and the external circuits or devices.

Princeton and Harvard Architecture

It is well known that in any microprocessor or microcontroller based system, the programs (instructions) and data are stored in the memory. Moreover, instructions and data are transferred from the memory to the processing unit for the corresponding execution (fig. 2.3).

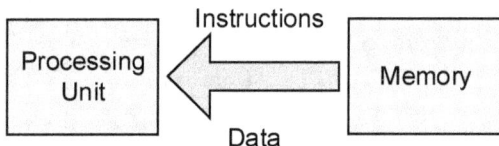

Figure 2.3 Transfer from the memory

The memory organization as well as the communication method with the processing unit affect directly the instruction execution time due to the fact that the transfer process introduces delays. On the other hand, if multiple buses are used (e.g. separate bus for addresses and data), then the above process is optimized regarding the instruction execution time (instructions and data are transferred simultaneously). From the first systems design, two basic architectures have been developed by the universities of Princeton and Harvard respectively. The main difference between the above architectures is the way that the memory is organized regarding the interconnection with the processing unit. In Princeton architecture (fig. 2.4), the program instructions and data are stored in the same memory, while the communication with the processing unit is implemented through the corresponding buses.

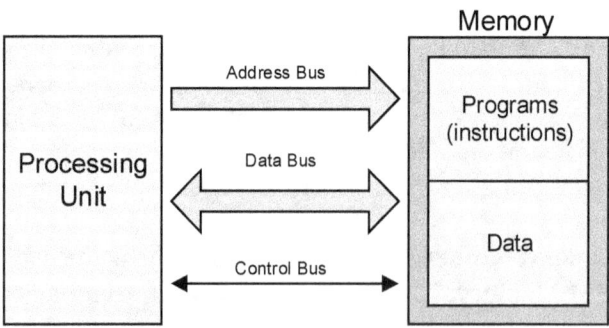

Figure 2.4 Princeton architecture

Based on the above architecture (fig. 2.4) two machine cycles are needed for executing one instruction because both the instruction and the data have to be transferred through the same bus. This architecture was selected during the 70's decade due to simplicity and the technological limitations of the era.

Due to the technological progress and the higher performance requirements, the Harvard architecture selected (fig. 2.5) where different memory is used for programs and data respectively. This practically means that separate buses can be used. Thus, operations can be implemented simultaneously. In other words, during the execution process of an instruction, the next instruction can be also transferred. With this method, the instruction execution requires only one machine cycle.

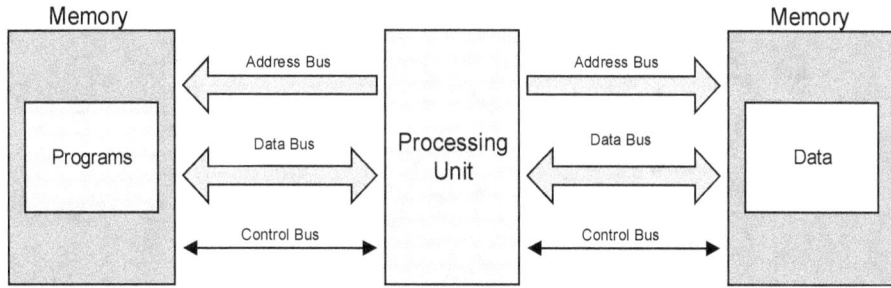

Figure 2.5 Harvard architecture

The AVR microcontrollers are based on the Harvard architecture.

The AVR memory system

As mentioned before, the AVR microcontrollers have separate memory areas for programs and data. Figure 2.6 shows the data memory which is implemented in a SRAM type memory as well as the program memory which is implemented in a Flash type memory. Moreover, EEPROM type memory is shown which holds data that have to be stored during the power off period. Based on the AVR model, the memory areas size may be different, while there is the option to use an additional external memory of SRAM type.

More precisely, the structure and the usage of the memory areas in figure 2.6, are analyzed as follows:

Data Memory
    Holds the program data and offers access to the general purpose registers of the microcontroller. Every memory location is 8bits.

    General purpose registers
        32 registers (R0 to R31) of 8bits are available to the programmer for developing applications.

    Input/Output (I/O) registers
        64 registers of 8bits have directly access to the microcontroller capabilities. For example, in order to send digital signals to an external device through the microcontroller ports, the corresponding data have to be stored in the correct I/O register. The same method

is applied also in other registers such as the timer control registers, etc.

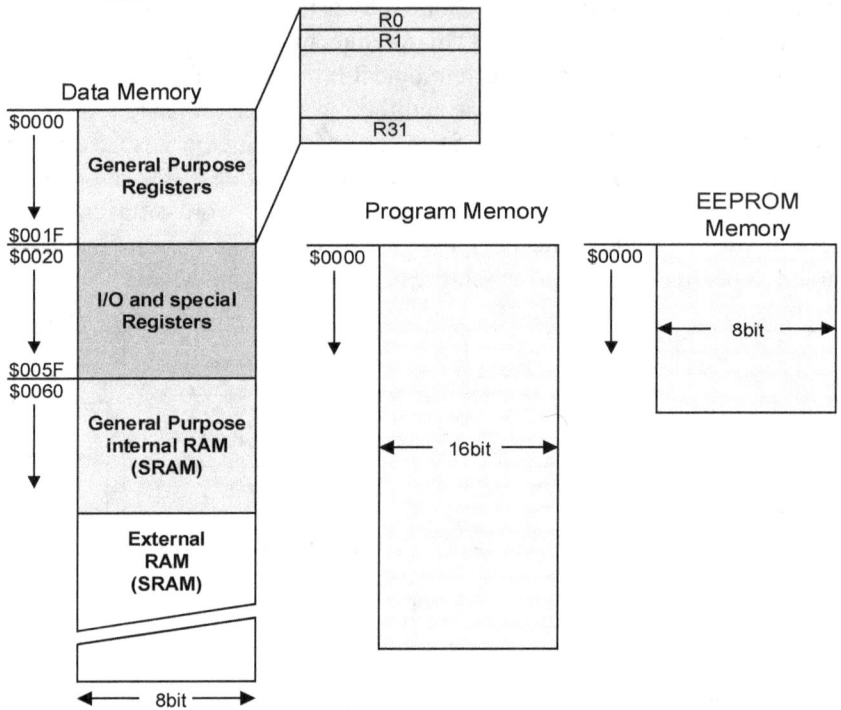

Figure 2.6 The AVR memory system

Internal RAM memory
   Is used for storing the program data.

External RAM memory
   This memory constitutes an external circuit and is used for storing the program data (for expanding the data memory).

Program memory
   Holds the applications code and it is different based on the microcontroller model. Each memory location is 16bits (2 bytes).

EEPROM memory
   This memory has limited storage capacity and is used for storing data that remain in the memory during the power off period. If some user settings have been stored in the data memory, then these settings are erased when the microcontroller is switched off. For holding the above settings during the power off period of the microcontroller, the EEPROM is used. Thus, in the next microcontroller initialization, the settings are available again to the user.

The I/O registers have directly access to the microcontroller hardware, while in some AVR models are more registers than the classical models. Thus, in the AVR models with more than 32 I/O pins there is also an additional area in the memory (extended I/O memory).

Figure 2.7 shows the data memory with the extended area for supporting more than 32 pins, while table 2.1 shows the memory size in various AVR models.

Table 2.1 Memory size in various AVR models

AVR model (Indicative)	RAM memory (main data memory)	Flash memory (program memory)	EEPROM memory (data memory)
ATmega48	512byte	4Kbyte	256byte
ATmega88	1Kbyte	8Kbyte	512byte
ATmega168	1Kbyte	16Kbyte	512byte
ATmega328	2Kbyte	32Kbyte	1Kbyte
ATmega32	2Kbyte	32Kbyte	1Kbyte
ATmega640	8Kbyte	64Kbyte	4Kbyte
ATmega1280	8Kbyte	128Kbyte	4Kbyte
ATmega2560	8Kbyte	256Kbyte	4Kbyte

Note
Based on the prefix, 1Kbyte=1x10^3 byte=1000byte. Due to the fact that addresses and data are expressed in binary, the corresponding bytes (1 byte = 8bits) have to represent this digits. Thus, 1Kbyte=1024 bytes, 2^{10}=1024 (closer to 1000). In practice, 8Kbytes means 8x1Kbytes=8x1024bytes=8192 bytes.

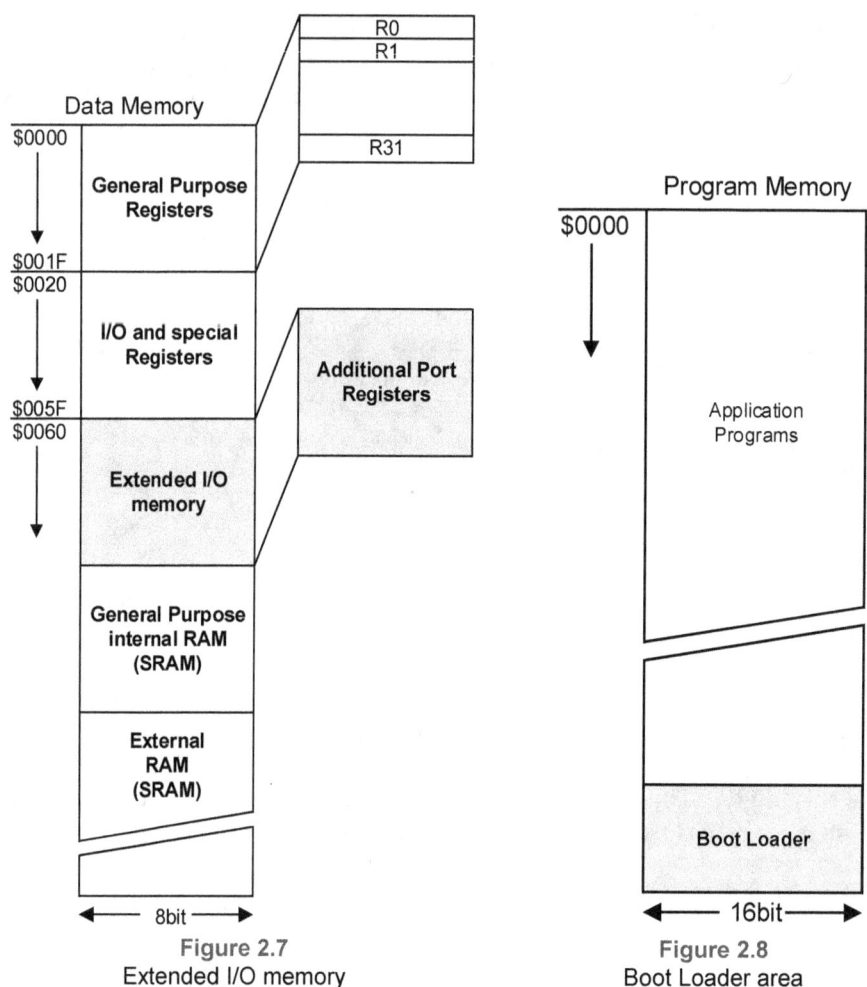

Figure 2.7
Extended I/O memory

Figure 2.8
Boot Loader area

In many AVR models, the program memory is divided in two areas (e.g. ATmega328), (a) applications program area and (b) boot area (Boot Loader). Figure 2.8 shows the two mentioned memory areas.

The Boot Loader supports the microcontroller reprogramming. Moreover, during the microcontroller activation the instruction execution can be started from this memory area and thus, the microcontroller behavior can be fully controlled.

General purpose registers

The AVR microcontrollers have 32 general purpose registers of 8bits that are mapped to specific addresses. Moreover, a group of 16bit registers exist. These registers are the X,Y,Z and are formed by grouping 8bit registers. These registers are also called "logical registers". Figure 2.9 shows the general purpose registers with the corresponding addresses as well as the logical registers X,Y and Z.

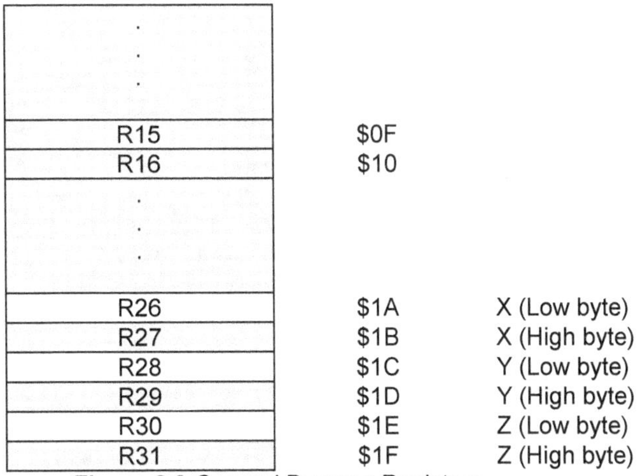

R15	$0F
R16	$10

R26	$1A	X (Low byte)
R27	$1B	X (High byte)
R28	$1C	Y (Low byte)
R29	$1D	Y (High byte)
R30	$1E	Z (Low byte)
R31	$1F	Z (High byte)

Figure 2.9 General Purpose Registers

Status Register (SREG)
The most important special register is the status register (SREG). The containing information of the SREG is used in order to investigate an ALU result, etc. Figure 2.10 shows the SREG structure.

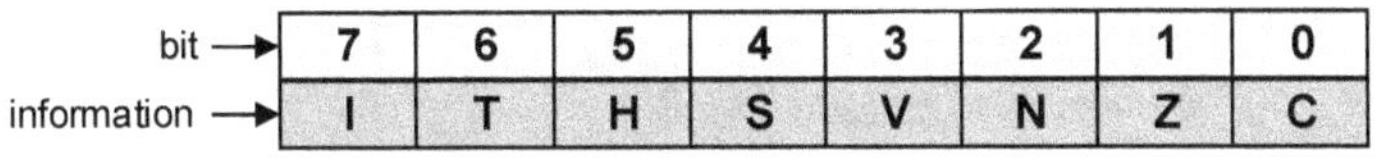

bit	7	6	5	4	3	2	1	0
information	I	T	H	S	V	N	Z	C

Figure 2.10 SREG bits

Table 2.2 shows the information which is indicated by each bit of the SREG.

Table 2.2 SREG bits

Bit	Description
I	Activating or deactivating the recognition of interrupts (events) from the outer environment
T	Is used for bit manipulation through the instructions BLD and BST
H	Carry which is produced from the third to the fourth bit of a number (e.g. in an ADD instruction)
S	Sign bit which is produced by an exclusive OR between the bits N and V
V	Overflow in a two's complement operation
N	Is activated when the MSB of a result is 1 (negative sign)
Z	Shows when the result of an arithmetic operation is zero
C	Carry (or borrow in an subtraction) which is produced in an arithmetical operation

I/O Registers
The exploitation of the architectural features of the AVR microcontroller (e.g. timers, communication with external devices), is mainly achieved through the I/O registers. The access to an I/O register can be performed by writing the corresponding address or the symbolic name. For example, through the register DDRB the data direction of the pins in port B is defined, while through the register PORTB, specific pins can be activated (e.g. to light up a LED).
Table 2.3 shows the available I/O registers of the AVR microcontroller.

Table 2.3 I/O registers (indicative)

Symbolic Name	Description
TWBR	Bit rate for the Serial Clock - SCL
TWSR	Two wire interface bus status
TWAR	Two wire address register for communication with an external slave device
TWDR	Data/Address Shift Register
ADCL	ADC (Analog to Digital Converter) Low byte
ADCH	ADC (Analog to Digital Converter) High byte
ADCSRA	ADC control and status register
ADMUX	ADC MUX register
ACSR	Analog comparator status register
UBRRL	UART baud rate (Low byte)
UBRRH	UART baud rate (High byte)
UCSRB	UART Control and status register B
UCSRA	UART Control and status register A
UDR	Indicates if the UART it is ready for reception
SPCR	SPI control register
SPSR	SPI status register
SPDR	SPI data register
PIND	Read (input) from the pins of port D
DDRD	Set data direction in port D
PORTD	Write (output) data in port D
PINC	Read (input) from the pins of port C
DDRC	Set data direction in port C
PORTC	Write (output) data in port C
PINB	Read (input) from the pins of port B
DDRB	Set data direction in port B
PORTB	Write (output) data in port B
PINA	Read (input) from the pins of port A
DDRA	Set data direction in port A
PORTA	Write (output) data in port A
EECR	EEPROM control register
EEDR	EEPROM data register
EEARL	EEPROM address register (Low byte)
EEARH	EEPROM address register (High byte)
WDTCR	WatchDog status register
ASSR	Asynchronous status register for timer/counter 1/2
OCR2	Register for comparing the value of timer/counter 2
TCNT2	Register for accessing the timer/counter 2
TCCR2	Control register of timer/counter 2
ICR1L	Low byte which is updated from timer/counter 1 when a specific event occurred
ICR1H	High byte which is updated from timer/counter 1 when a specific event occurred
OCR1BL	Low byte of register B which is always compared with the content of the timer/counter 1 in order to produce a specific event
OCR1BH	High byte of register B which is always compared with the content of the timer/counter 1 in order to produce a specific event
OCR1AL	Low byte of register A which is always compared with the content of the timer/counter 1 in order to produce a specific event
OCR1AH	High byte of register A which is always compared with the content of the timer/counter 1 in order to produce a specific event
TCNT1L	Low byte for direct access to timer/counter 1

TCNT1H	High byte for direct access to timer/counter 1
TCCR1B	Control register B of the timer/counter 1
TCCR1A	Control register A of the timer/counter 1
SFIOR	Special functions register (e.g. settings of the pull-up resistors)
OCDR	Debugging register
OSCCAL	Frequency register for calibrating the timing oscillator
TCNT0	Register for direct access to the timer/counter 0
TCCR0	Control register for the timer/counter 0
MCUCSR	Control and status register of the MCU (e.g. information about the cause of the recent reset)
MCUCR	MCU status register (e.g. interrupt event)
TWCR	Status register of serial bus
SPMCR	Control register for storing in the program memory
TIFR	Interrupt status register for the timer/counter
TIMSK	Interrupt mask register for the timer
GIFR	Interrupt status register (e.g. for the pins that support interrupts)
GICR	Interrupt control register (e.g. for the pins that support interrupts)
OCR0	Register which is always compared with the content of the timer/counter 0 in order to produce a specific event
SPL	Stack pointer (Low byte)
SPH	Stack pointer (High byte)
SREG	Status register

Note 1
Every I/O register is mapped to an address. This address is not the same among the AVR models. For example, the data direction register of port B (DDRB) is mapped to the address $04 in the AVR ATmega328 and to the address $17 in the AVR ATmega8515. Thus, in order to ensure the program functionality in different AVR models, only the symbolic name of the port is used. In order to activate the mapping of each I/O register address of the current AVR model to the corresponding symbolic name, special files (.INC) have to be embedded within the code.

Note 2
In previous figures, numbers of the hexadecimal system are used. The hexadecimal numbers are expressed with the prefix '$' or '0x'. For example, the number 0F is written as $0F or 0x0F. The symbol '$' is usually used for denoting addresses, while the symbol '0x' for denoting constant values (numbers).

EXERCISE SHEETS
Architectural features

1. Note every phrase as right or wrong

Phrase	Right	Wrong
A microcontroller supports general purpose applications	☐	☐

28 □ CHAPTER 2

An analog input recognizes voltages in the range [0-5]V	☐	☐
An MCU unit does not have common elements with a computer CPU	☐	☐
The Flash memory of the microcontroller loses the contents during the power off period	☐	☐
The Harvard architecture offers higher performance as compared to the Princeton architecture	☐	☐
All the program instructions are hosted inside the CPU	☐	☐
The MCU corresponds to the CPU	☐	☐
In Harvard architecture there is a double data bus and a single address bus	☐	☐
The EEPROM does not lose its contents during the power off period	☐	☐

2. Fill the missing text or words

2.1 The program counter points to the _____ of the next instruction that will be executed by the MCU/CPU.

2.2 The status register indicates mainly the _____ of _____ regarding the type _____ that is produced.

2.3 With Harvard architecture _____ machine cycle(s) is/are required for the _____ of every _____ as compared to the _____

3. If every instruction is 32bits long, how many memory locations of 16bit are required if the program consists of 75 instructions?

Calculation and results

4. If the program memory size is 20Kbyte and each memory location hosts a 16bit instruction, how many instructions can be stored in total?

Calculation and results

5. If the registers are 64, find the starting address of the I/O registers area.

Calculation and results

6. If the size of an external RAM is 16Kbyte and the starting address is $400, find the address of the last memory location.

Calculation and results

7. Search in the literature the features of the memory types SRAM, Flash and EEPROM and fill the following comparative table.

Feature	SRAM	Flash	EEPROM

8. The AVR ATmega32 has a RAM size of 2Kbyte. How many memory locations correspond?

Calculation and results

9. The RAM memory of a microcontroller consists of two areas of 17Kbyte and 11Kbyte respectively. How many are the memory locations in total?

Calculation and results

10. If R27=0x0F and R26=0xAA, find the content of the register X.

Answer

11. If Y=0xCBF0, find the content of the registers R28 and R29.

Answer

12. Fill the following table:

Memory size	Number of 8bit memory locations	Number of 16bit locations
2Kbytes		
	32768	
20Kbytes		
32Kbytes		
	16384	
	24576	

13. Write how many bits are required for storing the following numbers:
0xFF, 0xFA00, 0xC900, 0x100, 0xFFF, 0x00, 0x0200, 0x050, 0x04, 0x08, 0xFFFF

Number	Required bits
0xFF	
0xFA00	
0xC900	
0x100	
0xFFF	
0x00	
0x0200	
0x050	
0x04	
0x08	
0xFFFF	

14. Why is so important to use symbolic names for the I/O ports instead of the real addresses?

15. Write the disadvantage of using internal timer for the main clock.

16. Find ways to improve the analog input resolution of the microcontroller. Study the importance of the reference voltage and number of bits that are used for expressing the measurements.

17. Why the microcontroller uses a flash memory for storing the programs?

18. If the registers R0 and R1 are mapped to the addresses $00 and $01 respectively, find the corresponding addresses for the registers R19, R21 and R25.

Register	Address
R19	
R21	
R25	

19. A memory system in a microcontroller is organized in segments with size 16Kbytes+12Kbytes+20Kbytes (every segment is placed successively to the previous) respectively. Find the limit addresses of the segments (starting, ending) in decimal and hexadecimal system respectively.

Memory area	Limit addresses (decimal)	Limit addresses (hexadecimal)
16Kbyte	Starting = Ending =	Starting = Ending =
12Kbyte	Starting = Ending =	Starting = Ending =
20Kbyte	Starting = Ending =	Starting = Ending =

20. Search in the literature the pinout (port/pin diagram) of the ATmega328 and ATmega32 microcontrollers. Write in the following table the corresponding digital ports (name and number).

	Digital Port (e.g. A,B, etc)	Symbolic name (e.g. PA0, PA1, etc.)
ATmega328		

ATmega32		

21. Each location of the AVR data memory is _____

22. Each location of the AVR program memory is _____

23. Describe an application where the usage of the EEPROM memory is critical

24. Write the advantage of using an EEPROM memory instead of an EPROM memory (make a search in the literature).

25. Write application examples that are better to be developed using a microprocessor.

26. Search in the literature information, features and applications of the microprocessor Z80. Why an engineer will choose to develop an application using this microprocessor and not a modern one?

3 The basic Assembly instruction set of the microcontroller

Content-Goals
The basic application development tool is the available instruction set of the microcontroller. In this chapter each instruction is presented and analyzed with the required details for better understanding. Moreover, many examples are presented for better using of the instructions inside the code.

Chapter contents
3.1 Introduction
3.2 Loading instructions
3.3 Arithmetic instructions
3.4 Logical instructions
3.5 More arithmetic and other instructions
3.6 Bit manipulation
3.7 Developing complete and functional programs

3.1 Introduction

Despite the fact that the 8bit AVR microcontrollers are based on the RISC architecture, a rich instruction set is supported and many addressing modes as compared to other microcontrollers are available. The above features in combination with Harvard architecture (different data and program memory), offer satisfying performance and flexibility regarding the programs development. For better understanding of the first indicative programs and the operation philosophy of the AVR microcontrollers, some selected instructions will be analyzed in the next sections.

In assembly programming is useful to view also the data in other arithmetic systems for better understanding and for having a clue how the data are viewed by the microcontroller. It must be noticed that almost every instruction affects the content of a register which may affect the microcontroller behavior at physical level.

For example, in a case of port manipulation where every bit may correspond to a physical pin, the binary number representation is absolutely necessary. Figure 3.1 shows how the registers content will be presented in the next sections, while the used arithmetic system will be shown on the right side of the numerical value (content).

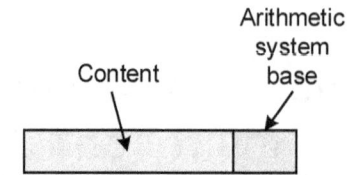

Figure 3.1 Register content presentation

Argument symbols

In the instruction presentation, arguments that represent registers, integer values and memory addresses are used. The argument symbols are shown in the table 3.1.

Table 3.1 Argument symbols

Symbol	Description
Rd	Destination register
Rs	Source register
k	Data (integer value)
addr	Constant address

3.2 Loading instructions

The addressing is referred to the data manipulation methods that involves load and transfer using registers, data and memory locations. The available methods for defining the source and the data destination, corresponds to the supported addressing modes of the microcontroller.

LDI – Direct addressing – Load an integer value to a register
For loading an integer value to a register, the following instruction is used:

`LDI Rd, k`

with $d \in [16,31]$, $k \in [0,255]$

The operation Rd=k will be performed, where the first argument is the destination. Figure 3.2 shows the value `k` which is loaded as a new content of the register `Rd`, while at the same time this value is also stored in the memory location where the register is mapped.

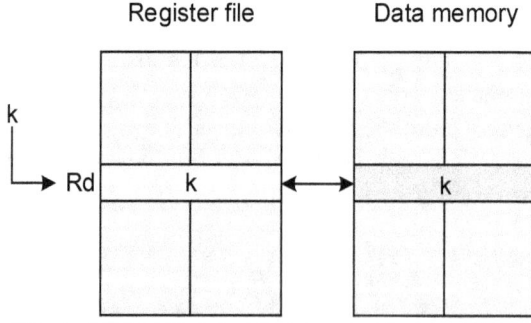

Figure 3.2 Loading an integer value to a register

Example

```
LDI R18,$0C        ;Loading the hexadecimal number 0C
                   ;in the register R18
LDI R18,0x0C       ;Loading the hexadecimal number 0C
                   ;in the register R18
LDI R18,12         ;Loading the decimal number 12
                   ;in the register R18
LDI R18,0b00001100 ;Loading the binary number 00001100
                   ;in the register R18
```

Register content after the instruction execution:

| 0C | 16 | 12 | 10 | 00001100 | 2 |

As shown in figure 3.3, the arithmetic value is the content of the specific register as well as the content of the memory location (data memory) where the register is mapped.

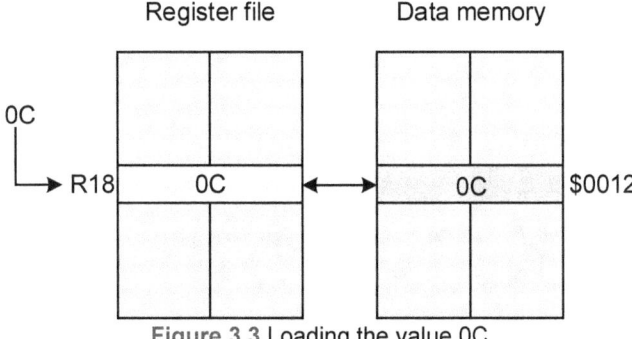

Figure 3.3 Loading the value 0C

Important: The LDI instruction does not work with the registers R0-R15.

Note
Arguments are adapted to current instruction specifications. For example, the instruction LDI uses the arguments Rd and k. The expression $d \in [16,31]$ means that, the d can take the values 16 to 31. In other words, the argument can be written as R16, R17, etc, in order to use the corresponding register. Similarly, the expression $k \in [0,255]$ means that, the k is an integer value in the range 0 to 255 (in hexadecimal 0x00 to 0xFF).

MOV – Direct addressing – Loading a register from a register
The loading instruction from register to register requires two registers as arguments. The loading operatin is performed with the instruction

```
MOV Rd,Rs
```
with $d,s \in [0,31]$

The instruction performs the operation Rd=Rs.

Example

It is assumed that R20=05 and the following instruction is executed:

```
MOV R17,R20
```

Figure 3.4 shows the instruction operation regarding the content of the registers and the data memory.

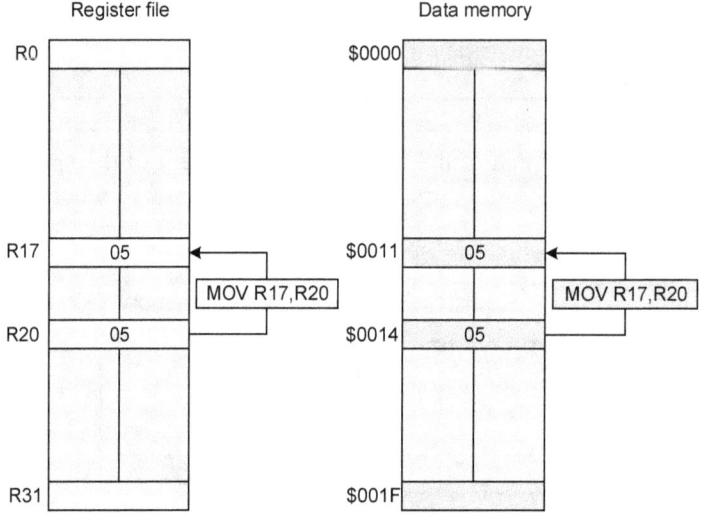

Figure 3.4 Loading a register from a register

Register content before instruction execution:

```
R17      ?     16      ?     10      ?         2
R20      05    16      05    10      00000101  2
```

```
MOV R17,R20
```

Register content after instruction execution:

```
R17      05    16      05    10      00000101  2
R20      05    16      05    10      00000101  2
```

Note
The symbol '?' means any content.

LDS – Direct addressing – Loading a register from a memory location
This instruction loads to a selected register the content of a memory location. For this operation the instruction form is:

```
LDS Rd,addr
```

with $d \in [0,31]$, addr $\in [0, 0xFFFF]$

The instruction implements the operation Rd=[addr]

Example
```
LDS R1,$201
```
As shown in figure 3.5, the content of the memory location with address $0201 (content 08) is loaded in the register R1.

Note
Usually, when an instruction uses an address, the prefix $ is used and not the 0x which is used for defining integer values. Using the above prefix, is more easy for the code to be read regarding the used addresses.

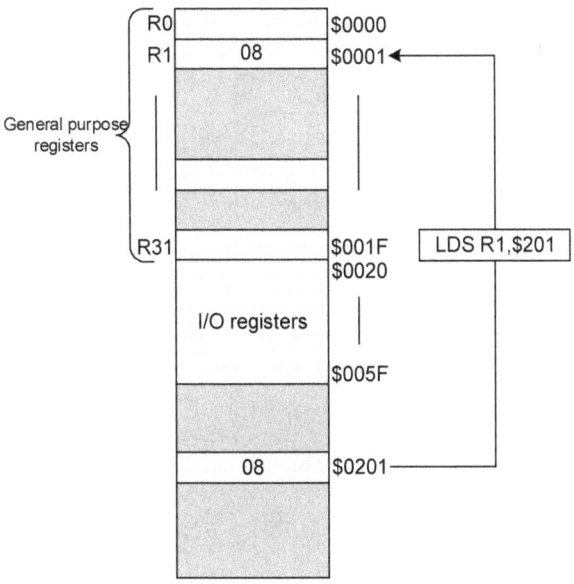

Figure 3.5 Loading a register from a memory location

STS – Direct addressing – Loading a memory location from a register
With this instruction the content of a register is stored in a specific memory location. The instruction arguments consists of the destination address and the source register.

```
STS addr,Rs
```

with addr ∈ [0,0xFFFF], s ∈ [0,31]

The address must exists in the microcontroller model that is used.
The above instruction performs the operation [addr]=Rs

Example
```
STS $201,R1
```

As shown in figure 3.6, the content of the register R1 (05) is stored in the memory location with address $0201.

Due to the fact that the registers are mapped to memory addresses, the address choice may have multiple operations. For example, the PORTC is mapped to the address $35 (for a specific AVR model). Thus, the storage to this memory location affects the physical PORTC.

The corresponding instruction will have the same form as the previously mentioned.

```
STS $35,R1   ;The data source can be any register
```
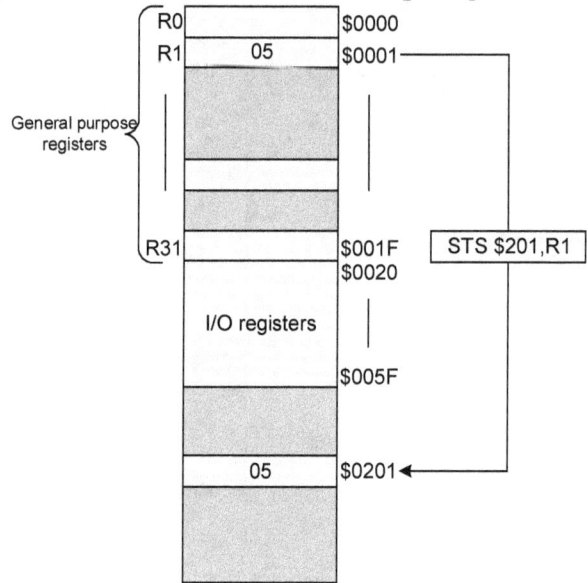

Figure 3.6 Loading a memory location from a register

The registers X,Y,Z – Indirect addressing
The 16bit registers X,Y and Z exist within the register file. These registers are not purely 16bits but they are formed by using 8bit registers. Thus, they are also called logical registers. As mentioned in previous chapter, in order to express an address, more than 8bits are required. That means that the 8bit registers (R1 to R31) can not be used for the memory access. Figure 3.7 shows the general purpose registers of the AVR.

In the figure 3.7 the register pairs R26:R27 (16 bit register X), R28:R29 (16 bit register Y) and R30:R31 (16 bit register Z) are also shown.

R26	$001A	X (Low order byte)
R27	$001B	X (High order byte)
R28	$001C	Y (Low order byte)
R29	$001D	Y (High order byte)
R30	$001E	Z (Low order byte)
R31	$001F	Z (High order byte)

Figure 3.7 General purpose registers

Figure 3.8 shows the organization of the "double" registers from the high to the low order byte regarding the registers R26 to R31.

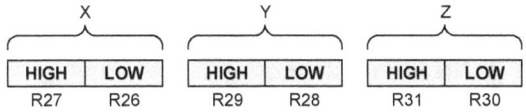

Figure 3.8 16bit registers organization

MOVW – Loading 16bit data in the register memory area

In many cases, there is a need for loading 16bits of data. This happens when calculations have to be performed with more than 8bits or addresses have to be manipulated. Due to the fact that every register is 8bits, the 16bits will be formed by using register pairs.

The instruction

```
MOVW Rd+1:Rd, Rs+1:Rs
```
with d,s ϵ {0,2,....,30}

loads in the pair Rd+1:Rd the 16bits data which are stored in the registers Rs+1 and Rs respectively.

Example

```
MOVW R1:R0, R21:R20
```

As shown in figure 3.9, the value 0x020A which is stored in the pair R21:R20, will be loaded in the pair R1:R0 respectively (R1=R21=02 and R0=R20=0A).

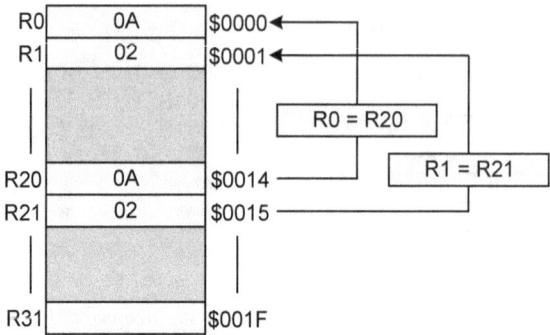

Figure 3.9 Loading 16bit data from a register pair to a register pair

Memory management through registers
The memory management (loading data from and to memory) is based:

(a) on 16bits addresses
(b) on the parametrized memory access for processing multiple memory locations

For met the above requirements, a 16bits register must be used. On the other hand, for managing multiple memory locations, a flexible memory access method has to be used. As mentioned before, each of the registers X, Y and Z is mapped in two memory locations in the data memory. Thus, they can be used for storing addresses. Moreover, the content of those registers can be automatically increased for accessing successive memory locations.

LD – Loading data from a memory location to a register
The instruction LD supports memory data management as follows:

<u>With register X</u>
```
LD Rd,X        ;Loading the contents of the address X
               ;in the register Rd
               ;d ∈ [0,31]
LD Rd,X+       ;Loading the contents of the address X
               ;in the register Rd and increment (after)
               ;of the X content by 1 (next address)
               ;d ∈ [0,31]
LD Rd,-X       ;Decrement of the X content by 1 and Loading (after)
               ;the contents of the address X in the register Rd
               ;d ∈ [0,31]
```

<u>With register Y</u>
```
LD Rd,Y        ;Loading the contents of the address Y
               ;in the register Rd
               ;d ∈ [0,31]
LD Rd,Y+       ;Loading the contents of the address Y
               ;in the register Rd and increment (after)
               ;of the Y content by 1 (next address)
```

	;d ∈ [0,31]
LD Rd,-Y	;Decrement of the Y content by 1 and Loading (after) ;the contents of the address Y in the register Rd ;d ∈ [0,31]
LDD Rd,Y+q	;Loading the contents of the address Y+q ;in the register Rd ;d ∈ [0,31]

With register Z

LD Rd,Z	;Loading the contents of the address Z ;in the register Rd ;d ∈ [0,31]
LD Rd,Z+	;Loading the contents of the address Z ;in the register Rd and increment (after) ;of the Z content by 1 (next address) ;d ∈ [0,31]
LD Rd,-Z	;Decrement of the Z content by 1 and Loading (after) ;the contents of the address Z in the register Rd ;d ∈ [0,31]
LDD Rd,Z+q	;Loading the contents of the address Z+q ;in the register Rd ;d ∈ [0,31]

Example 1 – Loading successively the contents of the memory locations $0200 and $0201 in the register R21(the previous contents are lost)

```
LDI R27,02    ;Load the high byte of the address
              ;($0200, the value 02) in the high byte of X
LDI R26,00    ;Load the low byte of the address ($0200,
              ;the value 00) in the low byte of X
LD R21,X+     ;Load the contents of the address $0200 to R21
              ;Increment the X content (after)
LD R21,X      ;Load the contents of the address $0201 to R21
```

The operation of the above instructions is analyzed as follows:

LDI R27,02
LDI R26,00

R27	02	16	02	10	00000010	2
R26	00	16	00	10	00000000	2

With the above instructions, the new content of register X will be:

X	0200	16	512	10	0000001000000000	2

LD R21,X+

As shown in figure 3.10, the pair R27:R26 forms the value 0x0200 within the register X. Thus, the value 0x07 (the content of the address $0200) is loaded in R21.

44 ☐ CHAPTER 3

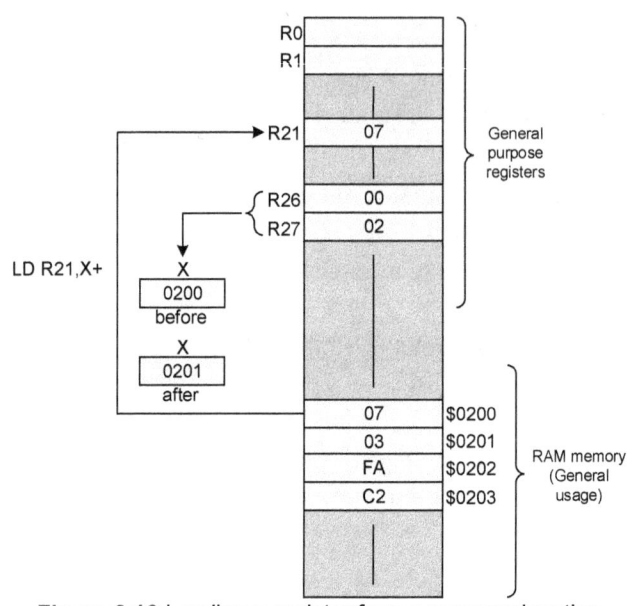

Figure 3.10 Loading a register from a memory location

After the instruction execution, the content of the register X has been increased by 1.

| X | 0201 | 16 | 513 | 10 | 0000001000000001 | 2 |

```
LD R21,X
```

The above instruction, loads the contents of the address $0201 (0x03) in the register R21.

Example 2 - Loading successively the contents of the memory locations $0200, $0201 and $0202 in the register R21 (the previous contents are lost).

```
LDI R27,02          ;Load the high byte
                    ;of the address ($0200, the value 02)
                    ;in the high byte of X

LDI R26,00          ;Load the low byte
                    ;of the address ($0200, the value 00)
                    ;in the low byte of X

LD R21,X+           ;Load the contents of the address X
                    ;in R21
                    ;Increment the X content (after)

LD R21,X+           ;Load the contents of the address X
                    ;in R21
                    ;Increment the X content (after)

LD R21,X            ;Load the contents of the address X
                    ;in R21
```

The results of the above load instructions in R21 are shown in figure 3.11.

The basic Assembly instruction set of the microcontroller   45

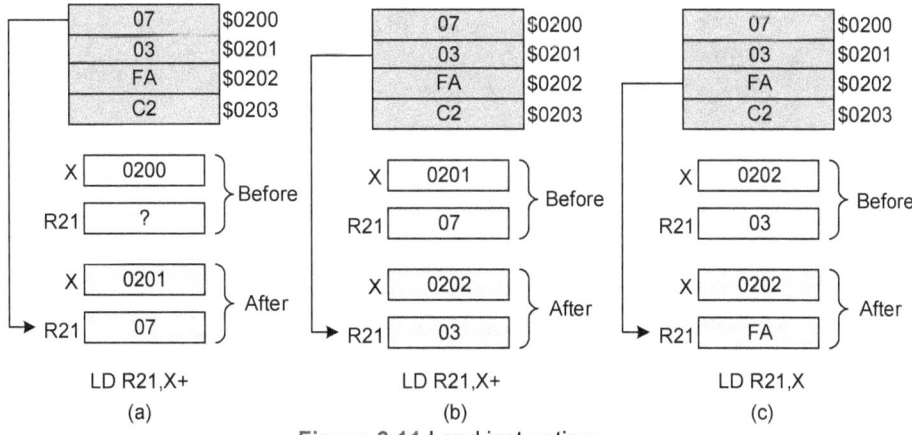

Figure 3.11 Load instructions

ST – Loading data from a register to a memory location
The instruction ST supports the data management of the registers as follows:

With register X
ST X,Rs	;Loading the Rs content to the memory location X ;d ∈ [0,31]
ST X+,Rs	;Loading the Rs content to the memory location X ;and increment of the X content by 1 (after) ;(next address) ;d ∈ [0,31]
ST -X,Rs	;Decrement of the X content by 1 (before) and loading the Rs ;content to the memory location X ;d ∈ [0,31]

With register Y
ST Y,Rs	;Loading the Rs content to the memory location Y ;d ∈ [0,31]
ST Y+,Rs	;Loading the Rs content to the memory location Y ;and increment of the Y content by 1 (after) ;(next address) ;d ∈ [0,31]
ST -Y,Rs	;Decrement of the Y content by 1 (before) and loading the Rs ;content to the memory location Y ;d ∈ [0,31]
STD Y+q,Rs	;Loading the Rs content to the memory location Y+q ;d ∈ [0,31]

With register Z
ST Z,Rs	; Loading the Rs content to the memory location Z ;d ∈ [0,31]
ST Z+,Rs	;Loading the Rs content to the memory location Z

46 □ CHAPTER 3

```
                    ;and increment of the Z content by 1 (after)
                    ;(next address)
                    ;d ∈ [0,31]
ST -Z,Rs            ;Decrement of the Z content by 1 (before) and loading the Rs
                    ;content to the memory location Z
                    ;d ∈ [0,31]
STD Z+q,Rs          ;Loading the Rs content to the memory location Z+q
                    ;d ∈ [0,31]
```

Example 1 – Loading the content of R21 to the address $0200

```
LDI R27,02          ;Load the high byte of the address
                    ;($0200, the value 02) in the high byte of X
LDI R26,00          ;Load the low byte of the address
                    ;($0200, the value 00) in the low byte of X
ST X,R21            ;Load the content of R21
                    ;to the address X ($0200)
```

The operation of the above instructions is analyzed as follows:
LDI R27,02
LDI R26,00

| R27 | 02 | 16 | 02 | 10 | 00000010 | 2 |
| R26 | 00 | 16 | 00 | 10 | 00000000 | 2 |

With the above instructions, the content of the register X will be

| X | 0200 | 16 | 512 | 10 | 0000001000000000 | 2 |

It is assumed that the content of the register R21, is

| R21 | 0F | 16 | 15 | 10 | 00001111 | 2 |

ST X,R21

As shown in figure 3.12, the pair R27:R26 forms the value 0x0200 in the register X. Thus, with the above instruction, the value 0x0F (content of the register R21) is loaded (store) to the memory location $0200.

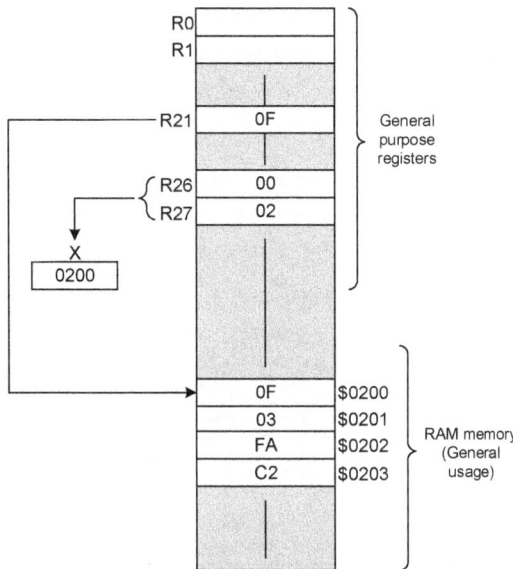

Figure 3.12 Loading a register content to a memory location

Example 2 - Loading successively the content of the register R21 to the memory locations $0200, $0201 and $0202 (the previous contents are lost).

```
LDI R27,02    ;Load the high byte of the address
              ;($0200, the value 02) in the high byte of X
LDI R26,00    ;Load the low byte of the address
              ;($0200, the value 00) in the low byte of X
ST  X+,R21    ;Load R21 to the memory location $0200
              ;X increment by 1
ST  X+,R21    ;Load R21 to the memory location $0201
              ;X increment by 1
ST  X,R21     ;Load R21 to the memory location $0202
```

It is assumed that the content of the register R21, is:

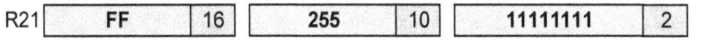

The results of the loading instructions of the register R21 to the memory locations are shown in figure 3.13.

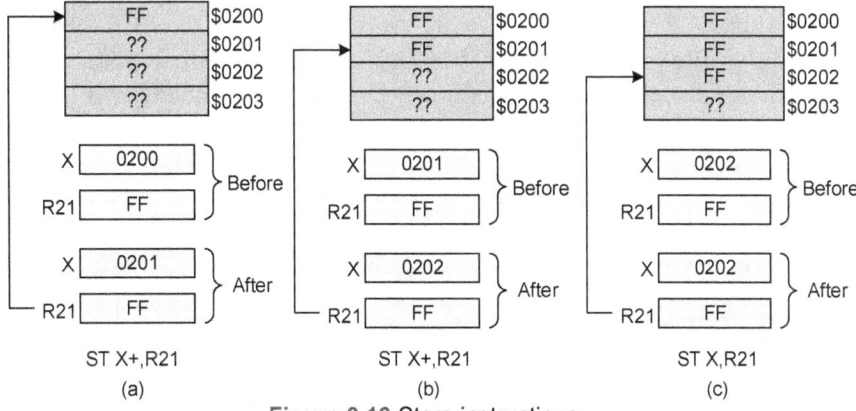

Figure 3.13 Store instructions

LPM – Loading from the program memory

The instruction `LPM` loads one byte from the program memory to a general purspose register. More precisely, the `LPM` can be used in three different forms:

`LPM`

Loads in the register `R0` one byte from the program memory address which is stored in the register `Z`.

`LPM Rd, Z`

Loads in the register `Rd` ($d \in [0,31]$) one byte from the program memory address which is stored in the register `Z`.

`LPM Rd, Z+`

Loads in the register `Rd` ($d \in [0,31]$) one byte from the program memory address which is stored in the register `Z`, while the content of the register `Z` is increased by one in order to point to the next address.

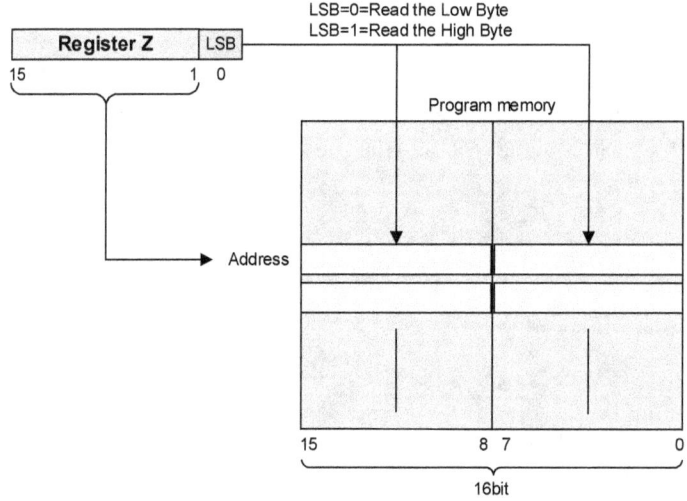

Figure 3.14 LPM though register Z

The Z is a 16bits register, but for defining the address of the program memory, only the 15 most significant bits are used (15 to 1). If all the 16bits are used for the address, then there is no place for the information regarding which of the two bytes of the program memory location will be read (the destination register is 8bits). Thus, the bits 15 to 1 are used for defining the address, while the bit 0 (LSB-least significant bit) defines which of the two bytes of the selected program memory location will be read. Figure 3.14 shows how the bits of the register Z are used through the instruction LPM. Moreover, when 15bits are used for defining the address, 2^{15} locations or 32K words can be accessed (every program memory location is 16bits long). On the other hand, if the microcontroller has a larger memory, then the instruction LPM introduces limitations. For that reason, the instruction ELPM which uses 24bit for reading from the program memory is used. More precisely, the pair RAMPZ:Z is used (the RAMPZ is an I/O register of 8bits). But in this case also, the LSB of the register pair is used in order to define which of the two bytes of the program memory will be read.

Data in the program memory
In many cases there is a need to declare data which will be stay unchanged (even after a reset or a power off condition) in order to be available always. This is achieved when those data are stored in the program memory (flash) and not in data memory.

(a) data declaration
With the directive .DB followed by the desired bytes, the required data sequence is declared. For example, the corresponding bytes for displaying digits on the seven segment display units (will be presented in next chapter) are declared within the code (inside the program memory), as follows:

```
digits:  .DB 0x3F,0x06,0x5B,0x4F,0x66,0x6D,0x7D,0x07,0x7F,0x6F
```

where digits, *is the symbolic address of the memory area where the data declaration starts*

(b) data size in relation with the program memory
The value which corresponds to each digit (of the above sequence) is expressed with one byte. On the other hand, every memory location in the program memory is 16bits long (2 bytes) as compared to the data memory where each location is 8bits long. This means that only using the address for reading of one byte is not enough. This issue can be also emphasized with the fact that the destination register is only 8bits long. Thus, which of the two bytes from the program memory location will be read in order to be stored in a general purpose register has to be selected (fig. 3.15).

(c) reading from the program memory using the corresponding addresses
For loading or reading from the program memory, special instructions are used. In this section, the instruction LPM will be used for reading from the program memory. The instruction LPM has to be used carefully in order to get the correct data from the program memory. For better understanding of the above operation it is assumed that one byte will be read from the program memory location with address 5.

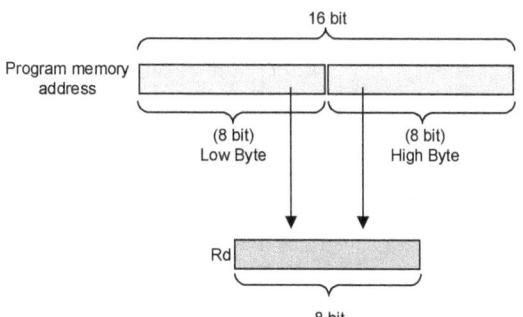

Figure 3.15 Loading one byte from the program memory

Using the 16bits of the register Z, the address 5 will be expressed as

0000000000000101

If the content of the register Z is used in the above form, then a byte will be read from the address 2 and not from 5. This happens because when the LPM is executed, the least significant (LSB) bit is used for selecting which of the two bytes from the program memory location will be read. This means that the rest of the bits (15bits) will be used for the address and thus, the final address is 2 (using 15bits). For solving this issue, a left shift of the initial content of the register Z is required (multiplication by 2). A left shift of the binary number 0000000000000101 (selected address), will produced the number 0000000000001010 (10 in decimal). In other words, for accessing the address 5 within the program memory, the value 10 has to be loaded in the register Z (the LSB bit is now available for selecting which of the two bytes will be read). The following examples show how the content of register Z is used in order to read data from the program memory starting from the address digits.

Example 1

In this example, the methods LOW and HIGH will be used in order to load the low and high bytes of the address digits in the register Z. At the same time, a multiplication by 2 is performed for preparing the content of the register Z before using the instruction LPM. After the Z initialization, the corresponding content can be increased for reading the next byte.

```
LDI     ZL,LOW(digits*2)       ;Preparing ZL (shift by multiplication)
LDI     ZH,HIGH(digits*2)      ;Preparing ZH (shift by multiplication)

LPM                            ;Read from the program memory
                               ;and store in R0
ADIW    ZL,1                   ;Increase Z for reading the
                               ;next byte
;Data array (byte sequence)
digits: .DB 0x3F,0x06,0x5B,0x4F,0x66,0x6D,0x7D,0x07,0x7F,0x6F
```

Example 2

In this case, the method << will be used for performing the left shift. Moreover, the content of the ZL is increased through a general purpose register (R19) by using an addition instruction. The above increment is used for reading the next byte from the array digits.

```
LDI     ZL,LOW(digits<<1)      ;Preparing ZL
LDI     ZH,HIGH(digits<<1)     ;Preparing ZH
ADD     ZL,R19                 ;Update ZL before the read from the
                               ;program memory
                               ;The size limitation of the ZL
                               ;must be taken in account
LPM                            ;Read from program memory

;Data array (byte sequence)
digits: .DB 0x3F,0x06,0x5B,0x4F,0x66,0x6D,0x7D,0x07,0x7F,0x6F
```

Table 3.2 summarizes the data load instructions that have been presented so far.

Table 3.2 Load instructions

Instruction	Description	Operation	Bits of the SREG that are affected
MOV Rd,Rs	Loading from register to register (8bit) d,s ∈ [0,31]	Rd=Rs	-
MOVW Rd,Rs	Loading from register to register (16bit) d,s ∈ {0,2,...,30}	Rd+1:Rd=Rs+1:Rs	-
LDI Rd,k	Loading an integer to a register (integer value) d ∈ [16,31], k ∈ [0,255]	Rd=k	-
LD Rd,X	Loading (indirect) from a memory location (pointed by X) to a register d ∈ [0,31]	Rd=(X)	-
LD Rd,X+	Loading (indirect) from a memory location (pointed by X) to a register and X increment by 1 (after) d ∈ [0,31]	Rd=(X) X=X+1	-
LD Rd,-X	X decrement by 1 (before) and Loading (indirect) from a memory location (pointed by X) to a register d ∈ [0,31]	X=X-1 Rd=(X)	-
LD Rd,Y	Loading (indirect) from a memory location (pointed by Y) to a register d ∈ [0,31]	Rd=(Y)	-
LD Rd,Y+	Loading (indirect) from a memory location (pointed by Y) to a register and Y increment by 1 (after) d ∈ [0,31]	Rd=(Y) Y=Y+1	-
LD Rd,-Y	Y decrement by 1 (before) and Loading (indirect) from a memory location (pointed by Y) to a register d ∈ [0,31]	Y=Y-1 Rd=(Y)	-
LDD Rd,Y+q	Loading (indirect) from a memory location (pointed by Y+q) to a register (q is an integer value) d ∈ [0,31]	Rd=(Y+q)	-
LD Rd,Z	Loading (indirect) from a memory location (pointed by Z) to a register d ∈ [0,31]	Rd=(Z)	-
LD Rd,Z+	Loading (indirect) from a memory location (pointed by Z) to a register and Z increment by 1 (after) d ∈ [0,31]	Rd=(Z) Z=Z+1	-
LD Rd,-Z	Z decrement by 1 (before) and Loading (indirect)	Z=Z-1 Rd=(Z)	-

Instruction	Description	Operation
	from a memory location (pointed by Z) to a register d ∈ [0,31]	
LDD Rd,Z+q	Loading (indirect) from a memory location (pointed by Z+q) to a register (q is an integer value) d ∈ [0,31]	Rd=(Z+q)
LDS Rd,addr	Loading (direct) a register from a memory location (RAM) d ∈ [0,31]	Rd=(addr)
ST X,Rs	Loading (indirect) from a register to a memory location (pointed by X) s ∈ [0,31]	(X)=Rs
ST X+,Rs	Loading (indirect) from a register to a memory location (pointed by X) and X increment by 1 (after) s ∈ [0,31]	(X)=Rs X=X+1
ST −X,Rs	X decrement by 1 (before) and Loading (indirect) a register to a memory location (pointed by X) s ∈ [0,31]	X=X−1 (X)=Rs
ST Y,Rs	Loading (indirect) from a register to a memory location (pointed by Y) s ∈ [0,31]	(Y)=Rs
ST Y+,Rs	Loading (indirect) from a register to a memory location (pointed by Y) and Y increment by 1 (after) s ∈ [0,31]	(Y)=Rs Y=Y+1
ST −Y,Rs	Y decrement by 1 (before) and Loading (indirect) a register to a memory location (pointed by Y) s ∈ [0,31]	Y=Y−1 (Y)=Rs
STD Y+q,Rs	Loading (indirect) from a register to a memory location (pointed by Y+q), q is an integer value s ∈ [0,31]	(Y+q)=Rs
ST Z,Rs	Loading (indirect) from a register to a memory location (pointed by Z) s ∈ [0,31]	(Z)=Rs
ST Z+,Rs	Loading (indirect) from a register to a memory location (pointed by Z) and Z increment by 1 (after) s ∈ [0,31]	(Z)=Rs Z=Z+1
ST −Z,Rs	Z decrement by 1 (before) and Loading (indirect) a register to a memory location (pointed by Z) s ∈ [0,31]	Z=Z−1 (Z)=Rs

STD Z+q,Rs	Loading (indirect) from a register to a memory location (pointed by Z+q), q is an integer value s ∈ [0,31]	(Y+q)=Rs	-
STS addr,Rs	Loading (direct) a memory location (RAM) from a register s ∈ [0,31]	(addr)=Rs	-

3.3 Arithmetic instructions

The AVR microcontroller has instructions for supporting addition, subtraction and multiplication. Moreover, does not support the division operation and thus, the corresponding algorithm and code has to be developed by the programmer. This algorithm can be based on successive subtractions or on other approaches. The code development regarding the division operation emphasizes the significance of the theoretical knowledge. On the other hand, the arithmetic or logical shift can be used for performing multiplication or division by 2. Finally, it must be noticed that the AVR supports also the primitive logical operations (AND, OR, NOT).

ADD – Adding two registers (without carry)

The instruction

```
ADD Rd,Rs
```
with $d,s \in [0,31]$

performs the operation Rd=Rd+Rs, while the bits H, S, V, N, Z and C of the register SREG can be affected.

Example

```
ADD R18,R20
```

Before the instruction execution, it is assumed the following register contents (R18 and R20).

| R18 | 0A | 16 | 10 | 10 | 00001010 | 2 |
| R20 | 0B | 16 | 11 | 10 | 00001011 | 2 |

With the instruction ADD (fig. 3.16), the content of the two registers is added and the result is stored in the first argument (R18, destination register).

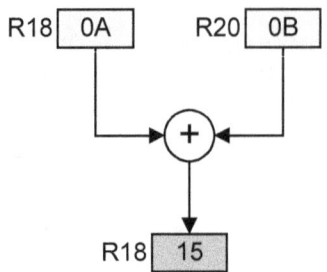

Figure 3.16 Adding two registers

As shown below, the result is 15 (0x15).

| R18 | 15 | 16 | 21 | 10 | 00010101 | 2 |

In contrast to the loading instructions, the arithmetic and logical instructions affect some bits of the status register (SREG). Thus, it is very important to know which of the above bits are affected in order to perform correctly the corresponding operations without losing any data.

As mentioned in previous chapter, the AVR microcontroller has a special unit which is called ALU (Arithmetic and Logic Unit). The ALU accepts two arithmetic arguments and performs the corresponding operation between them, while some bits of the SREG are affected from the result.

ADC – Adding two registers with carry

The instruction

```
ADC Rd,Rs
```
with $d,s \in [0,31]$

performs the operation Rd=Rd+Rs+C, where C is the carry value of the SREG. In many cases, the result is 9bits long and cannot be stored in a 8bits register. Thus, the 9th bit which is the produced carry has to be combined with the rest 8bits for forming the final result (number).

Example (adding two 8bits numbers)

In this example, two 8bit numbers will be added and the result will be stored in the memory locations $7F (High Byte) and $7E (Low Byte).
It is assumed the following content of the registers R18 and R19:

| R18 | FE | 16 | 254 | 10 | 11111110 | 2 |
| R19 | 07 | 16 | 07 | 10 | 00000111 | 2 |

In this example, the first instruction that will be executed is

```
ADD R18,R19
```

Figure 3.17 shows how the ALU is involved in the addition of the two registers. If a carry is produced, then this is represented by a specific bit within the SREG.

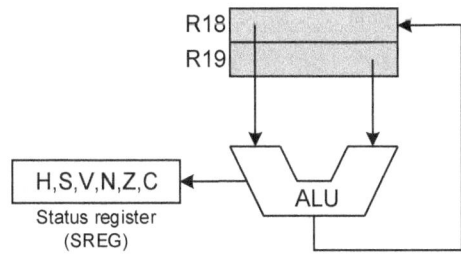

Figure 3.17 The ALU role in the addition operation

Figure 3.18 shows the numbers addition in two phases: (1) number transferring in the ALU and (2) performing addition and transfering the result.

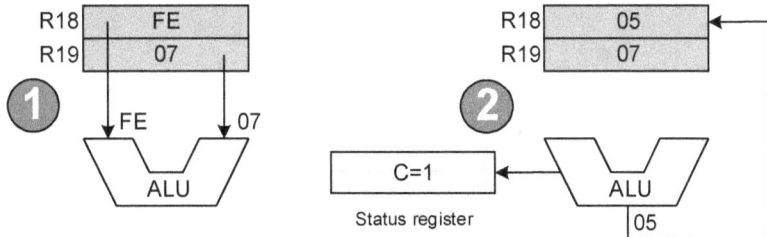

Figure 3.18 Numbers addition in two phases

It must be noticed that the phases of the figure 3.18 are part of the whole execution operation which require only one machine cycle in total. The addition of the numbers 0xFE and 0x07, gives as a result the number 0x105 which cannot be represented only with 8bits. In other words, the register R18 is not enough for storing the result despite the fact the is used as a first argument in the instruction ADD. For better understanding of the above issue and for facing effectively the problem, the above addition will be studied using the binary system. Based on figure 3.19, a carry is produced after the most significant bit (MSB) of the two numbers. Thus, the first 8bits of the result which represent the number 05 are stored in the register R18. The carry (9th bit of the result), is the activated bit C of the status register (SREG).

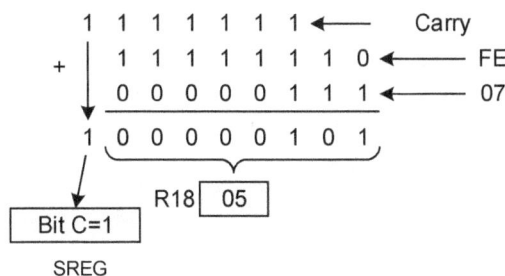

Figure 3.19 Addition in the binary system

As shown below, the sum consists of 9bits in total.

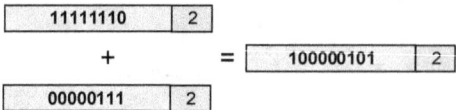

Due to the fact that the result will be stored inside registers (or memory locations), two registers will be required, while the final result will be read correctly using the two registers as a pair.

Based on the above, the register R18 will hold the eight least significant bits (LSBs) of the result, while the 9th bit will be read from the SREG. Figure 3.20 shows how the produced carry will be used. The most significant byte will be derived from the carry which has to be loaded in a register.

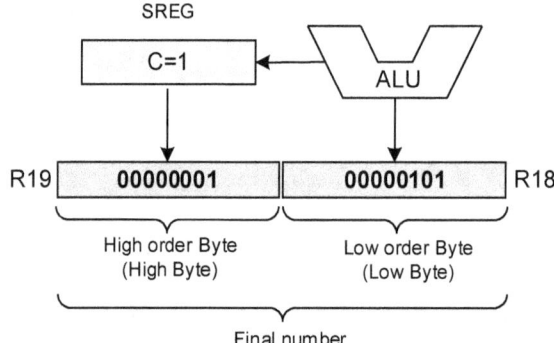

Figure 3.20 Forming the final number by using the carry

Figure 3.20 shows that the register R18 will hold the least significant bits that have been produced by the addition, while the register R19 will hold the carry. Initially, the register R19 has to be initialized due to the fact that the instruction ADC adds the carry in the register R19. Thus, the following instructions will be executed:

```
LDI R19,0x00       ;R19=0x00
ADC R19,R19        ;R19=R19+C
```

Now, the register R19 holds the most significant byte of the result. The final number is stored in the pair R19:R18. At the beginning of the example, is mentioned that the final number (result) will be stored in the memory locations $7F and $7E respectively. For achieving that, the following instructions will be used:

```
STS $7F,R19        ;($7F)=R19
STS $7E,R18        ;($7E)=R18
```

The final code is as follows:
```
LDI R18,0xFE   ;Initial value in R18
LDI R19,0x07   ;Initial value in R19
ADD R18,R19    ;R18=R18+R19
LDI R19,0x00   ;R19=0x00
ADC R19,R19    ;R19=R19+C
```

```
STS $7F,R19    ;($7F)=R19
STS $7E,R18    ;($7E)=R18
```

Example (multiple precision addition - 16bit, 2bytes result)

In this case, two numbers of 2bytes (16bit) each will be added. The result will be stored in the memory locations $0201 and $0200 (High/Low). It is assumed that the numbers 0x0127 and 0x04B0 will be added. The number 0x0127 is expressed using 2bytes as:

00000001 00100111$_{(2)}$ (0x0127)

while the number 0x04B0 is expressed as:

00000100 10110000$_{(2)}$ (0x04B0)

The eight first bits (from the left) represent the most significant byte (High byte), while the rest of the bits represent the least significant byte (Low byte). Table 3.3 shows the corresponding bytes for the above numbers.

Table 3.3 Number bytes

	High and Low byte		
0x0127		0x04B0	
High byte	Low byte	High byte	Low byte
00000001	00100111	00000100	10110000

The addition will be performed separately for the High and Low bytes (fig. 3.21).

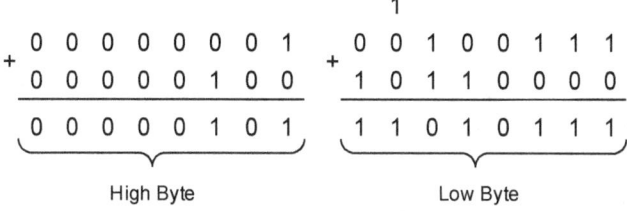

Figure 3.21 Addition operation

Combining the results of the figure 3.21, the number 0000010111010111 is derived (the number 0x05D7 which represents the sum 0x0127+0x04B0).
If no carry is produced (especially in the addition of the Low Bytes), then the final number is just the merge of High και Low bytes which are read from left to right.
If a carry is produced from the addition of the low bytes, then will be added together with the high bytes. For understanding the above carry manipulation, the addition 0x01CE+0x05FD will be performed. These two numbers are expressed in the binary system as follows:

0x01CE = 00000001 11001110$_{(2)}$
0x05FD = 00000101 11111101$_{(2)}$

Figure 3.22 shows the addition of the High and Low bytes of the two numbers. It must be noticed the carry addition to the High bytes. As mentioned before, the final number is produced by merging the resulted High and Low bytes.

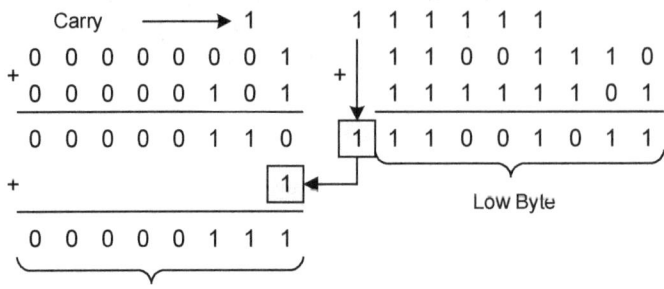

Figure 3.22 High, Low byte addition

For implementing the above addition, the High and Low bytes of the numbers will be stored in the register pairs R20:R19 and R18:R17 respectively. The code for the above addition is as follows:

```
                           ;Loading of the 0x01CE using two 8bit
                           ;numbers
LDI  R19,0b11001110        ;Load the Low Byte
LDI  R20,0b00000001        ;Load the High Byte

                           ;Loading of the 0x05FD using two 8bit
                           ;numbers
LDI  R17,0b11111101        ;Load the Low Byte
LDI  R18,0b00000101        ;Load the High Byte

ADD  R17,R19               ;Adding the Low Bytes
LDI  R16,0x00              ;Initialize R16 for storing the carry
ADC  R16,R16               ;Load carry in R16

ADD  R18,R20               ;Adding the High Bytes
ADC  R18,R16               ;Adding the carry
                           ;from the low Bytes

STS  $0201,R18             ;Store High Byte in $0201
STS  $0200,R17             ;Store Low Byte in $0200
```

Example (multiple precision addition – 16bit, 3bytes result)
In this example, two numbers of 2 bytes (16bit) each will be added, while the result cannot be expressed only with 2 bytes.
Thus, 3 bytes are needed for the result. This happens because a carry is produced from the High bytes addition.
In this example, the numbers 0xED59 and 0xC7D7 will be added. These numbers are expressed in the binary system as follows:

0xED59 = 1110110101011001₍₂₎
0xC7D7 = 11000111110101111₍₂₎

Figure 3.23 shows the addition of the two numbers as well as the result which consists of 3 bytes ($B_2B_1B_0$).

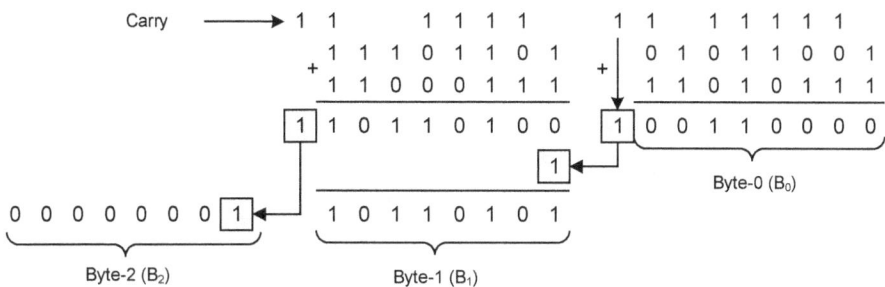

Figure 3.23 Number addition with 3 bytes result

As shown in figure 3.23, a carry is produced from the Low bytes addition (B_0). This carry is used in the High bytes addition for forming the sum which represents the most significant byte (B_1). Finally, the produced carry cannot be stored in the second byte and thus, a third byte is used (B_2).

The final number is derived by putting the three bytes sorted by significance. Thus, the number is

000000001 10110101 00110000₍₂₎
The implementation code is as follows:

```
LDI R19,0b01011001      ;Load the Low Byte
LDI R20,0b11101101      ;Load the High Byte
```
The value 0x59 (Low byte) is loaded in the register R19, while the value 0xED (High byte) is loaded in the register R20

```
LDI R17,0b11010111      ;Load the Low Byte
LDI R18,0b11000111      ;Load the High Byte
```
The value 0xD7 (Low byte) is loaded in the register R17, while the value 0xC7 (High byte) is loaded in the register R18

```
ADD R17,R19
```
After the addition of the Low bytes (0xD7+0x59), the sum (0x30) is stored in the register R17, while a carry is produced (the C Flag of the SREG is activated)

```
LDI R16,0x00
```
The register R16 is cleared in order to hold the carry after the ADC instruction execution

```
ADC R16,R16
```

The operation R16=R16+R16+C is performed and the new content of the register R16 is only the carry

```
ADD R18,R20
```
After the High bytes addition (0xC7+0xED), the sum (0xB4) is stored in the register R18, while a carry is produced (the C Flag of the SREG is activated)

```
LDI R20,0x00
ADC R20,R20
```
Before adding the produced carry from the Low bytes, the produced carry from the High bytes addition is temporarily stored. This is necessary because if the carry of the Low bytes is added before any other operation, then the previous carry will cleared due to the fact that the C flag of the SREG is affected after any arithmetic operation. The carry is stored in the register R20.

```
ADD R18,R16
```
Now the carry from the Low bytes is added (was stored in R16)

```
STS $0202,R20
STS $0201,R18
STS $0200,R17
```
After the arithmetic operations, the memory locations $0202 and $0200 will contain the final sum from the most to the least significant byte. The code for the above operations is as follows:

```
                        ;Loading of the 0xED59 using two 8bit
                        ;numbers
LDI R19,0b01011001      ;Load the Low Byte
LDI R20,0b11101101      ;Load the High Byte
                        ;Loading of the 0xC7D7 using two 8bit
                        ;numbers
LDI R17,0b11010111      ;Load the Low Byte
LDI R18,0b11000111      ;Load the High Byte
ADD R17,R19             ;Adding the Low Bytes
LDI R16,0x00            ;Clear R16 for storing
                        ;the carry
ADC R16,R16             ;Load the carry in R16
                        ;from the Low Bytes sum
ADD R18,R20             ;Adding the High Bytes
LDI R20,0x00            ;Clear R20 for storing
                        ;the carry
ADC R20,R20             ;Load the carry in R20
                        ;from the High bytes sum
ADD R18,R16             ;Adding the carry from the
                        ;Low Bytes to High Bytes
STS $0202,R20           ;$0202=byte2
STS $0201,R18           ;$0201=byte1
STS $0200,R17           ;$0200=byte0
```

ADIW – Adding an integer number to a 16bit registers pair
The 16bit numbers manipulation is performed using registers pairs. That means that the number which forms using the registers pair has to be faced as a unified number.

For example, if the number 0xFF is stored in the register R30 (Low byte of the Z register) and the number 0x01 is stored in the register R31 (High byte of the register Z), then the unified 16bit number which is produced is 0x01FF. If the number 1 is added to the above number, then the result will be 0x0200. This result (0x200) must be produced from an ADD instruction which faces the registers pair content as a unified 16bit number. Such an instruction is the `ADIW` which is expressed as follows:

```
ADIW Rd,k
```

with $d \in \{24,26,28,30\}$, $k \in [0,63]$

This instruction performs the operation Rd+1:Rd=Rd+1:Rd+k. In other words, the `ADIW` instruction adds the integer value k in the registers pair Rd+1:Rd where the corresponding content will be faced as a unified 16bit number. The d belongs to the range $\{24,26,28,30\}$ and thus the register pairs R25:R24, R27:R26, R29:R28 and R31:R30 are formed. On the other hand, the integer value k belongs to the range [0,63].

Example
It is initially assumed that the register Z (R31:R30) holds the value 0x01FF (R31=0x01, R30=0xFF, fig. 3.24). In this example the number 1 will be added three times to the above number using the instruction `ADIW`. This operation will be implemented as follows:

```
LDI R30,0xFF
LDI R31,0x01
```

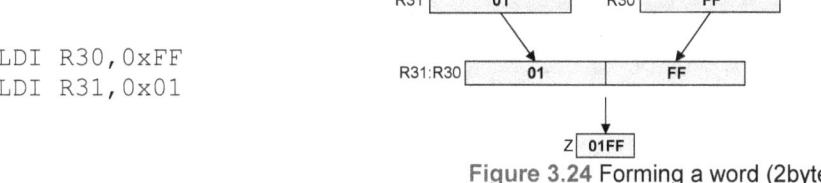

Figure 3.24 Forming a word (2byte)

The above instructions (`LDI`) load the values 0x01 and 0xFF in the registers R31 and R30 respectively and practically Z= 0x01FF.

```
ADIW R30,1
```
| 01FF | +1= | 0200 |

```
ADIW R30,1
```
| 0200 | +1= | 0201 |

```
ADIW R30,1
```
| 0201 | +1= | 0202 |

The above instructions (`ADIW`) add the number 1 to the unified 16bit number. Thus, the successive values are 0x0200, 0x0201 and 0x0202. If the instruction `ADD R30,R16` (R16=1) is used instead of `ADIW`, then the addition 0xFF+1 will be performed, which will has as a result the clearance of the register R30 and the activation of the C flag (SREG) because 0xFF+1=0x100.

SUB – Subtracting two registers (without carry)

The instruction

```
SUB Rd,Rs
```

with d,s ∈ [0,31]

performs the operation Rd=Rd-Rs, while the result may affect the status of the bits H, S, V, N, Z and C of the status register (SREG).

Example 1
In this example, the content of the register R20 is less than the content of the register R18. Thus, the result is positive.

```
SUB R18,R20  ;R18=R18-R20
```

Before the instruction execution it is assumed that the registers R18 and R20 have the following content:

| R18 | 0F | 16 | 15 | 10 | 00001111 | 2 |
| R20 | 0A | 16 | 10 | 10 | 00001010 | 2 |

With the instruction SUB (fig. 3.25), a subtraction is performed between two registers, while the result is stored in the first register (first argument).

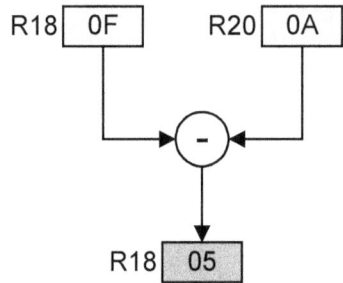

Figure 3.25 Subtraction

The result of the above subtraction is 05 (0x05).

| R18 | 05 | 16 | 05 | 10 | 00000101 | 2 |

The AVR performs the subtraction using the two's complement. In this example, the two's complement is not yet presented due to the fact that R20 is less than R18 and the result is the expected. However, the subtraction has been performed by using the two's complement. In practice, for calculating the arithmetic operation R18-R20, the microcontroller performs the operation R18+(-R20). In other words, the two's complement of R20 is added to R18. The sign and the magnitude of the final number is derived from the result. The two's complement is used to calculate the opposite of a given number. For calculating the two's complement of a number, the bits are inverted while the number 1 is added. For the subtraction 0x0F-0x0A, the arithmetic operation 0x0F+(-0x0A) will be performed. For better understanding, the above numbers will be written in the binary form.

$0x0F = 00001111_{(2)}$
$0x0A = 00001010_{(2)}$
$-0x0A = (11110101+1)_{(2)} = 11110110_{(2)}$

The calculation 0x0F+(-0x0A), will be performed as follows (fig. 3.26):

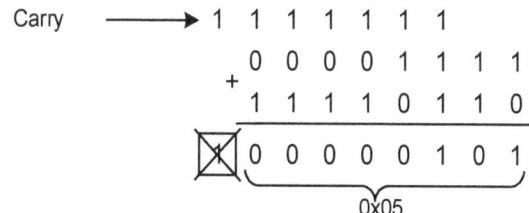

Figure 3.26 Addition (subtraction) using the two's complement

If a carry is produced, then it is ignored and the result is positive with a magnitude which is derived from the rest of the bits.

Example 2

In this example, the subtraction R20-R18 will be performed, where the result is negative. Initially it is assumed that the content of the registers R20 and R18 is

| R18 | 0F | 16 | 15 | 10 | 00001111 | 2 |
| R20 | 0A | 16 | 10 | 10 | 00001010 | 2 |

Thus, the arithmetic operation R20+(-R18) will be performed.

When the instruction `SUB R20, R18` is executed the new content of the register R20 will be

| R20 | FB | 16 | 251 | 10 | 11111011 | 2 |

For understanding the above result, the corresponding arithmetic operation will be analyzed in the binary system by calculating the 0x0A-0x0F, that is 0x0A+(-0x0F).

The representation of the above numbers in the binary form is:

$0x0A = 00001010_{(2)}$
$0x0F = 00001111_{(2)}$
$-0x0F = (11110000+1)_{(2)} = 11110001_{(2)}$

Thus, the arithmetic operation 0x0A+(-0x0F) will be performed in the binary system as follows (fig. 3.27):

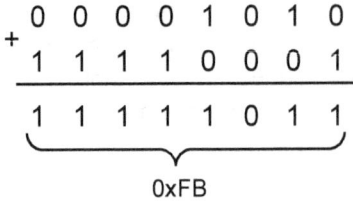

Figure 3.27 Addition (subtraction) using the two's complement

In the above calculation, no carry is produced and thus, the result is negative with a magnitude which is equal to the two's complement of the result. The magnitude of the result is produced by calculating the two's complement (invert the bits and add the number 1):

$(00000100+1)_{(2)}=(00000101)_{(2)}$

The magnitude of the number is 5, while the sign is already known. Thus, the calculation 0x0A-0x0F=-0x05 or $10_{(10)}-15_{(10)}=-5_{(10)}$ is verified.

SUBI – Subtracting an integer value from a register
The instruction

```
SUBI Rd,k
```
with $d \in [16,31]$, $k \in [0,255]$

performs the operation Rd=Rd-k.

Example

```
SUBI R20,0x0F
```

It is assumed that the content of the register R20 is

| R20 | AA | 16 | 170 | 10 | 10101010 | 2 |

When the instruction `SUBI R20,0x0F` is executed the new content of the register R20 will be

| R20 | 9B | 16 | 155 | 10 | 10011011 | 2 |

SBC – Subtracting two registers with carry
The instruction
```
SBC Rd,Rs
```
with $d,s \in [0,31]$

Performs the operation Rd=Rd-Rs-C, where C is the bit value (carry flag) of the register SREG.

Example

```
LDI R20,0xFF   ;Load the number 0xFF in R20
LDI R21,0x01   ;Load the number 0x01 in R21
ADD R20,R21    ;Addition R20=R20+R21
SBC R21,R21    ;R21=R21-R21-1
```

The addition 0xFF+0x01 will activate the carry flag of the SREG (C=1) and thus, with the instruction SBC the arithmetic operation 0x01-0x01-1 will be performed. Due to the fact that the result is -1, the final content of the register R21 will be 0xFF (-1 according to the two's complement).

SBCI – Subtracting an integer value and the carry from a register
The instruction

```
SBCI Rd,k
```

with $d \in [16,31]$, $k \in [0,255]$

subtracts from the register Rd, the integer value k, as well as the carry.

SBIW – Subtracting an integer value from a registers pair of 16bits
The instruction

```
SBIW Rd,k
```

with $d \in \{24,26,28,30\}$, $k \in [0,63]$

subtracts the integer value k from the registers pair Rd+1:Rd. Like the instruction ADIW, the goal of the above instruction is to update the destination register as a unified 16bit number.

3.4 Logical instructions

AND – Logical AND between registers
The instruction

```
AND Rd,Rs
```

with $d,s \in [0,31]$

performs the logical operation AND in the corresponding bits of the two registers and the result is stored in the register Rd.
Table 3.4 shows the truth table of the logical AND.

Table 3.4 Logical AND

A	B	A AND B
0	0	0
0	1	0
1	0	0
1	1	1

In the above table, the result of the logical operation AND using two hypothetical variables is presented.

Example
Assuming that R18=0xAC and R23=0x32, the following instruction will be executed:

```
AND R18,R23
```

The above logical operation will be executed as follows (fig. 3.28):

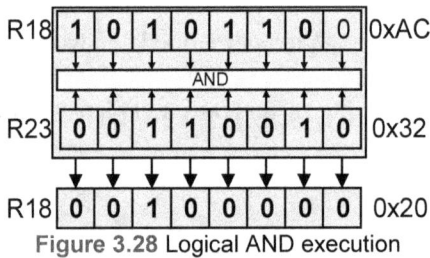

Figure 3.28 Logical AND execution

ANDI – Logical AND between register and integer value
The instruction

```
ANDI Rd,k
```
with d ϵ [16,31], k ϵ [0,255]

performs a logical AND in the corresponding bits of the register Rd and the bits which represent the integer value k. This instruction performs the same logical operation with the previous instruction, but the second argument is based directly to an integer value.

OR – Logical OR between registers
The instruction

```
OR Rd,Rs
```
with d,s ϵ [0,31]

performs a logical OR between the corresponding bits of the two registers and the result is stored in the register Rd. Table 3.5 shows the truth table for the logical OR.

Table 3.5 Logical OR

A	B	A OR B
0	0	0
0	1	1
1	0	1
1	1	1

In the above table, the result of the logical operation OR using two hypothetical variables is presented.

Example
Assuming that R18=0xAC and R23=0x32 the following instruction will be executed:

```
OR R18,R23
```

The above logical operation will be executed as follows (fig. 3.29):

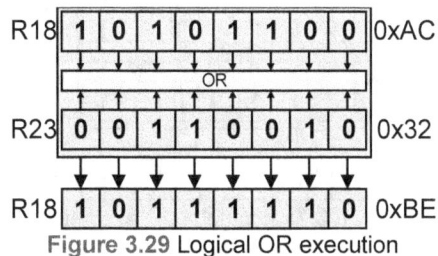

Figure 3.29 Logical OR execution

ORI – Logical OR between register and integer value
The instruction

```
ORI Rd,k
```
with $d \in [16,31]$, $k \in [0,255]$

performs a logical OR in the corresponding bits of the register Rd and the bits which represent the integer value k. This instruction performs the same logical operation with the previous instruction, but the second argument is based directly to an integer value.

EOR – Logical EXCLUSIVE OR (XOR) between registers
The instruction

```
EOR Rd,Rs
```
with $d,s \in [0,31]$

performs a logical XOR between the corresponding bits of the two registers and the result is stored in the register Rd. Table 3.6 shows the truth table for the logical XOR.

Table 3.6 Logical XOR

A	B	A XOR B
0	0	0
0	1	1
1	0	1
1	1	0

In the above table, the result of the logical operation XOR using two hypothetical variables is presented.

Example
Assuming that R18=0xAC and R23=0x32, the following instruction will be executed:
EOR R18,R23
The above operation will be performed as follows (fig. 3.30):

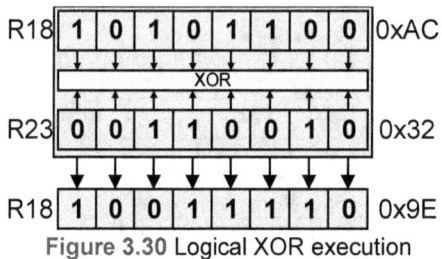

Figure 3.30 Logical XOR execution

COM – Logical NOT (one's complement) of a register content

The instruction
COM Rd
with $d \in [0,31]$

performs a logical NOT to the corresponding bits of the register Rd and the result is stored in the same Rd. Table 3.7 shows the truth table for the logical NOT.

Table 3.7 Logical NOT

A	NOT A
0	1
1	0

In the above table, the result of the logical operation NOT using a hypothetical variable is presented.

Example
Assuming that R18=0xAC, the following instruction will be executed:

```
COM R18
```

The above logical operation will be executed as follows (fig. 3.31):

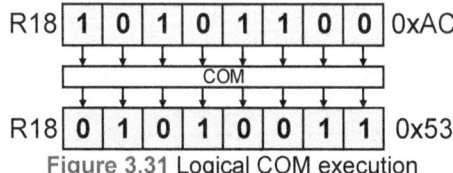

Figure 3.31 Logical COM execution

NEG – Calculating the two's complement of a register
The instruction

```
NEG Rd
```

with $d \in [0,31]$

forms the two's complement of the register Rd and the result is stored in the same register.

Example
Assuming that R18=0xAC the following instruction will be executed

```
NEG R18
```

As shown in the subtraction operation, for calculating the two's complement, the bits of the corresponding number have to be inverted while the number 1 will be added. As will be shown, the complement of the value 0xAC is 0x54 which now represents the new register content (fig. 3.32).

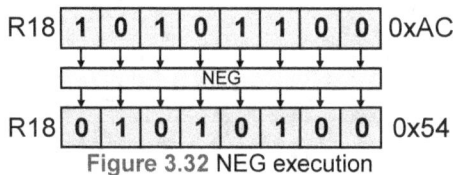

Figure 3.32 NEG execution

3.5 More arithmetic and other instructions

INC – Increasing a register content by 1
The instruction

```
INC Rd
```

with $d \in [0,31]$

performs the operation Rd=Rd+1, (adding an 1 in the content of the register Rd), while the result is stored in the same register.

Example

Assuming that R18=0xAC, the following instruction will be executed:

```
INC R18
```

Figure 3.33 shows the content of the register R18 before and after the above instruction execution.

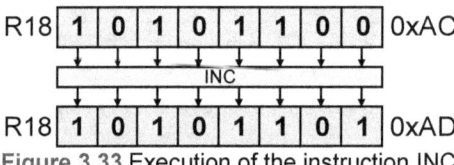

Figure 3.33 Execution of the instruction INC

DEC – Decrement a register content by 1

The instruction

```
DEC Rd
```
with $d \in [0,31]$

performs the operation Rd=Rd-1, (subtracting an 1 from the content of the register Rd), while the result is stored in the same register.

Example

Assuming that R18=0xAC, the following instruction will be executed:

```
DEC R18
```

The following figure (fig. 3.34) shows the content of the register R18 before and after the instruction execution.

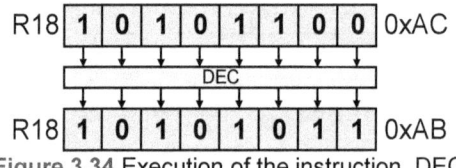

Figure 3.34 Execution of the instruction DEC

CLR – Clearing a register content

Many instructions exist only for helping the programmer in the program development process, but actually are implemented with more complicated instructions from the side of the microcontroller.
The instruction

```
CLR Rd
```
with d ∈ [0,31]

clears the content of the register Rd. Actually, an EXCLUSIVE OR (XOR) is performed between the contents of the register (the XOR is performed between the same two inputs). As a result, all the register bits are cleared, due to the fact that a logical XOR gives 1 only if the corresponding bits are different.

Example
In this example, the new content of the register R18 (fig. 3.35) will be shown after executing the following instruction:

```
CLR R18
```

It is assumed that the initial content of the register R18 is 0xAC.

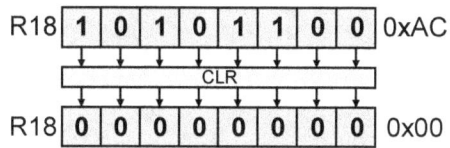

Figure 3.35 Execution of the instruction CLR

The same result is produced (fig. 3.36) if the following instruction is executed.
```
EOR R18,R18
```

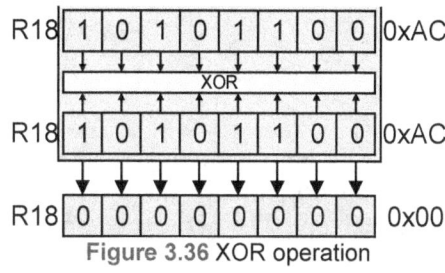

Figure 3.36 XOR operation

SER – Loading the maximum value to a register
The instruction

```
SER Rd
```
with d ∈ [16,31]

sets to all bits the value 1, that means that the new content of the register will be 255 or 0xFF, independently from the previous content.

Example

```
SER R20
```

The operation of the above instruction is shown in figure 3.37.

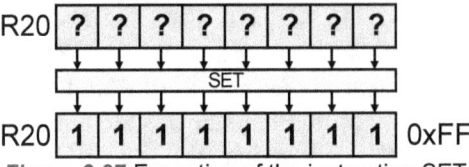

Figure 3.37 Execution of the instruction SET

MUL – Unsigned numbers multiplication through registers

The arithmetic multiplication is a simple operation especially when the two numbers are unsigned. That means that all the bits are used for the number magnitude. The multiplication arguments are 8bit long because of the registers. On the other hand, the result has to be stored correctly. It is obvious that the result may cannot be stored in a 8bit register. Thus, two registers (16bit long in total) have to be combined in order to store the result.
The instruction

```
MUL Rd,Rs
```

with $d,s \in [0,31]$

multiplies the content of the registers Rd and Rs and the result is stored in the registers pair R1:R0 as follows:

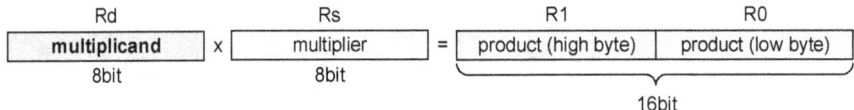

Example
In this example, the number 0xF0 will be multiplied with 0x7B, while the result will be stored in the registers pair R7:R6. Initially, the two numbers will be loaded in the registers R18 and R19 respectively. The above operation will be implemented as follows:

```
LDI R18,0xF0
```

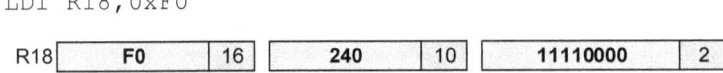

Loading the number 0xF0 in the register R18.

```
LDI R19,0x7B
```

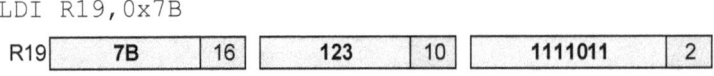

Loading the number 0x7B in the register R19.

```
MUL R18,R19
```

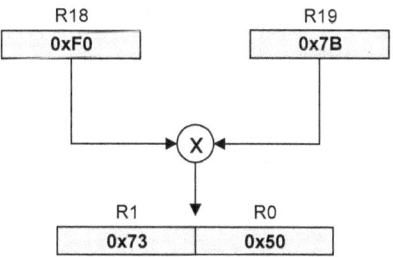

Figure 3.38 Multiplication

Multiplication 0xF0 * 0x7B and automatic storage of the result in the registers pair R1:R0 (fig. 3.38).

```
MOVW R6,R0
```

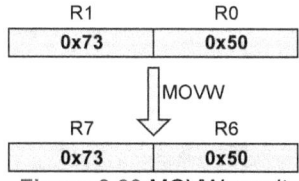

Figure 3.39 MOVW result

The result is 16bits and thus, the instruction MOVW will be used (fig. 3.39) in order to be transferred to a new registers pair.

The complete code follows:
```
LDI  R18,0xF0   ;Load the 0xF0 on R18
LDI  R19,0x7B   ;Load the 0x7B on R19
MUL  R18,R19    ;Multiplication (R1:R0=R18*R19)
MOVW R6,R0      ;Load the result (R7:R6=R1:R0)
```

MULS – Signed numbers multiplication through registers
The instruction

```
MULS Rd,Rs
```

with $d,s \in [16,31]$
multiplies the content of the registers Rd and Rs and the result is stored in the registers pair R1:R0 as follows:

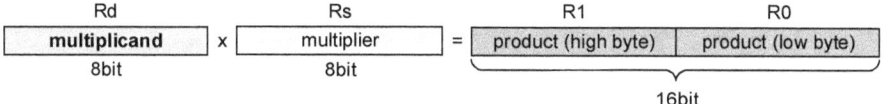

Example 1
The most important issue is the number sign manipulation. The instruction NEG can be used to form the two's complement. On the other hand, in the instruction MULS,

the microcontroller takes in account the sign of the numbers. In this example a positive number will be multiplied with a negative number. The negative number will be produced by applying the instruction NEG. The expected result will be negative. Thus, this result has to be analyzed. In the following program, the number -0x7F will be multiplied with 0x02. The -0x7F is produced by applying the instruction NEG to the number 0x7F. The result is stored in the registers pair R1:R0.

```
LDI R18,0x7F
```
| R18 | 7F | 16 | 127 | 10 | 01111111 | 2 |

```
LDI R19,0x02
```
| R19 | 02 | 16 | 02 | 10 | 00000010 | 2 |

```
NEG R18
```
| R18 | 81 | 16 | 129 | 10 | 10000001 | 2 |

```
MULS R18,R19
```
| R1 | FF | 16 | 255 | 10 | 11111111 | 2 |
| R0 | 02 | 16 | 02 | 10 | 00000010 | 2 |

Facing the registers pair R1:R0 as a unified 16bit number, the following number can be expressed:

$1111111100000010_{(2)}$

The negative sign is derived from the most significant bit of the number. Thus, the two's complement will be produced for calculating the magnitude of the number. The bits of the number are inverting and the number 1 is added, as follows:

$0000000011111101_{(2)}+1_{(2)}=0000000011111110_{(2)}$

The number $0000000011111110_{(2)}$ equals to $FE_{(16)}$ or $254_{(10)}$. Thus, the above arithmetic operation is verified.

Example 2

In this example, the instruction NEG will be used twice in the initial numbers. By applying the above operations, the resulted numbers are negative and the corresponding result will be positive.

```
LDI R18,0x7A
```
| R18 | 7A | 16 | 122 | 10 | 01111010 | 2 |

```
LDI R19,0x03
```
| R19 | 03 | 16 | 03 | 10 | 00000011 | 2 |

```
NEG R18
```
| R18 | 86 | 16 | 134 | 10 | 10000110 | 2 |

```
NEG R19
```

| R19 | FD | 16 | 253 | 10 | 11111101 | 2 |

MULS R18,R19

| R1 | 01 | 16 | 01 | 10 | 00000001 | 2 |
| R0 | 6E | 16 | 110 | 10 | 01101110 | 2 |

The content of R1 denotes that the result is positive with a 16bit magnitude which is stored in the registers pair R1:R0. Based on the above, the arithmetic operation - 0x7A*(-0x03)=0x016E or $122_{(10)}$ x $3_{(10)}=366_{(10)}$ is verified.

MULSU – Multiplication of a signed number with an unsigned number through registers

The instruction MULSU multiplies two numbers by combining a signed with an unsigned value.

The instruction

`MULSU Rd,Rs`

with $d,s \in [16,23]$

performs the operation R1:R0=Rd*Rs, which can be described by the following figure:

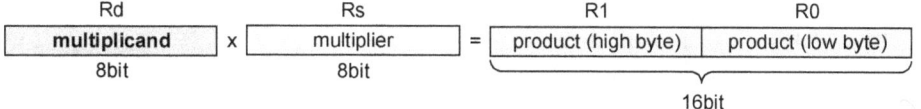

In the above arithmetic operation the register Rd contains the signed number, while the register Rs contains the unsinged number.

FMUL/FMULS – Multiplication of unsigned/signed fractioned numbers

The instruction FMUL/FMULS multiplies two unsigned/signed 8bit numbers and produces a 16bit product (with one left shift of the result).

The instruction

`FMUL Rd,Rs`

with $d,s \in [16,23]$

implements the operation R1:R0=Rd*Rs

If the (X.Y) represents a fractional number with X binary digits on the left side of the floating point and Y binary digits on the right side of the floating point, then the multiplication (X1.Y1)*(X2.Y2) will produce a result in the form ((X1+X2).(Y1+Y2)).

A similar instruction is the **FMULSU** which multiplies a signed with an unsigned number.

Table 3.8 summarizes the arithmetic and logical instructions.

Table 3.8 Arithmetic and Logical instructions

Instruction	Description	Operation	Bits of the SREG that are affected
ADD Rd,Rs	Add two registers (without carry) d,s ∈ [0,31]	Rd=Rd+Rs	H,S,V,N,Z,C
ADC Rd,Rs	Add two registers with carry d,s ∈ [0,31]	Rd=Rd+Rs+C	H,S,V,N,Z,C
ADIW Rd,k	Add an integer value to a 16bit registers pair d ∈ {24,26,28,30}, k ∈ [0,63]	Rd+1:Rd=Rd+1:Rd+k	S,V,N,Z,C
SUB Rd,Rs	Subtract two registers (without carry) d,s ∈ [0,31]	Rd=Rd-Rs	H,S,V,N,Z,C
SUBI Rd,k	Subtract an integer value from a register d ∈ [16,31], k ∈ [0,255]	Rd=Rd-k	H,S,V,N,Z,C
SBC Rd,Rs	Subtract two registers with carry d,s ∈ [0,31]	Rd=Rd+Rs-C	H,S,V,N,Z,C
SBCI Rd,k	Subtract an integer value and the carry from a register d ∈ [16,31], k ∈ [0,255]	Rd=Rd-k-C	H,S,V,N,Z,C
SBIW Rd,k	Subtract an integer value from a 16bit registers pair d ∈ {24,26,28,30}, k ∈ [0,63]	Rd+1:Rd=Rd+1:Rd-k	S,V,N,Z,C
AND Rd,Rs	Logical AND between two registers d,s ∈ [0,31]	Rd=Rd AND Rs	S,V=0,N,Z
ANDI Rd,k	Logical AND between a register and an integer value d ∈ [16,31], k ∈ [0,255]	Rd=Rd AND k	S,V=0,N,Z
OR Rd,Rs	Logical OR between registers d,s ∈ [0,31]	Rd=Rd OR Rs	S,V=0,N,Z
ORI Rd,k	Logical OR between a register and an integer value d ∈ [16,31], k ∈ [0,255]	Rd=Rd OR k	S,V=0,N,Z
EOR Rd,Rs	Logical EXCLUSIVE OR between two registers d,s ∈ [0,31]	Rd=Rd XOR Rs	S,V=0,N,Z
COM Rd	Logical NOT (one's complement) of a register content d ∈ [0,31]	Rd=0xFF-Rd	S,V=0,N,Z,C=1
NEG Rd	Two's complement calculation of a register content	Rd=0x00-Rd	H,S,V,N,Z,C

INC Rd	d ϵ [0,31] Increment the register content by 1	Rd=Rd+1	S,V,N,Z
DEC Rd	d ϵ [0,31] Decrement the register content by 1	Rd=Rd-1	S,V,N,Z
CLR Rd	d ϵ [0,31] Clear the register content	Rd=0x00	S=0,V=0,N=0,Z=1
SER	d ϵ [16,31] Load the maximum value in a register	Rd=0xFF	-
MUL Rd,Rs	d ϵ [0,31] Unsigned numbers multiplication through registers	R1:R0=Rd*Rs	Z,C
MULS Rd,Rs	d,s ϵ [0,31] Signed numbers multiplication through registers	R1:R0=Rd*Rs	Z,C
MULSU Rd,Rs	d,s ϵ [16,31] Signed/Unsigned numbers multiplication trhough registers	R1:R0=Rd*Rs	Z,C
FMUL	d,s ϵ [16,23] Unsigned fractioned numbers multiplication	R1:R0=Rd*Rs	Z,C
FMULS	d,s ϵ [16,23] Signed fractioned numbers multiplication	R1:R0=Rd*Rs	Z,C
FMULSU	d,s ϵ [16,23] Multiplication of a signed fractioned number with an unsigned fractioned number d,s ϵ [16,23]	R1:R0=Rd*Rs	Z,C

Note
Based on the AVR microcontroller model, some instructions maybe not supported. Thus, the corresponding AVR datasheet must be always taken in account.

3.6 Bit manipulation

The bit manipulation instructions control selected bits inside general or special purpose registers. The above manipulation can be also performed by logical instructions but this approach is more complex. The logical instructions perform also logical or arithmetic shift as will be presented.

LSL – Logical shift left
The instruction

```
LSL Rd
```
with d ϵ [0,31]

shifts to the left the content of the register Rd by one digit where the most significant bit (MSB) is stored in the carry bit (CF) of the SREG. This happens in order not to lose the MSB digit which is rejected from the register due to the shift operation. From the right side (LSB digit), the content of the register is filling with zeros. The above operation is analyzed as follows:

C=Rd(7)   Loading the carry from the bit 7 (MSB) of the register
Rd(n+1)=Rd(n)   The bit n is shifting to the next left position
Rd(0)=0   The bit 0 is loaded with zero

Example
Initially, the value 0x16 will be loaded in the register R20 and then, a logical shift left will be performed three times. Thus, the following instructions will be executed:

```
LDI R20,0x16    ;Load the initial value
LSL R20         ;1st shift
LSL R20         ;2nd shift
LSL R20         ;3rd shift
```

Figure 3.40 shows the phases of the multiple shift to the left that is implemented by the above code. The initial value is 0x16 (22 decimal).

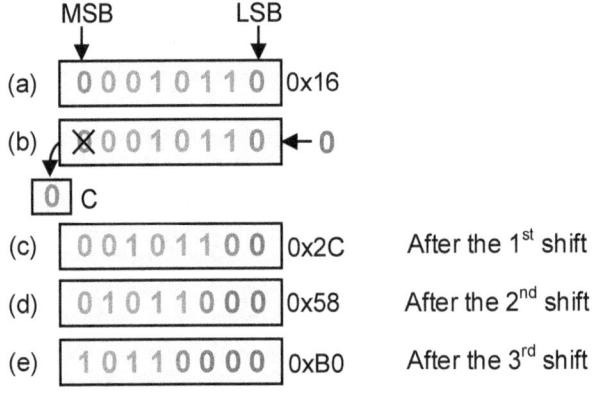

Figure 3.40 Logical shift left

The above phases (fig. 3.40) are analyzed as follows:

(a) the initial number is $0x16 = 22_{(10)}$
(b) in the first shift, the MSB digit is rejected (is stored in the bit C of the SREG), while the number is filled on the right with a zero
(c) after the first shift, the new content of the register is $0x2C = 44_{(10)}$
(d) after the second shift, the content of the register is $0x58 = 88_{(10)}$
(e) after the third (last) shift the content of the register will be $0xB0 = 176_{(10)}$

Table 3.9 shows the successive values of the register according to the above shifts.

Table 3.9 Results of the logical shifts (initial number 22)

	Initial number	1st shift	2nd shift	3rd shift
Hexadecimal	16	2C	58	B0
Decimal	22	44	88	176

Based on the table 3.9, in every shift operation the content of the register is multiplied by 2. The above operation is very important because the multiplication by 2 can be performed with a simple shift instruction.

LSR – Logical shift right
The instruction

```
LSR Rd
```

with $d \in [0,31]$

shifts to the right the content of the register Rd by one digit where the least significant bit (LSB) is stored in the carry bit (CF) of the SREG. This happens in order not to lose the LSB digit which is rejected from the register due to the shift operation. From the left side (MSB digit), the content of the register is filling with zeros. The above operation is analyzed as follows:

$C=Rd(0)$ Loading the carry from the bit 0 (LSB) of the register
$Rd(n)=Rd(n+1)$ The bit n is shifting to the next right position
$Rd(7)=0$ The bit 7 is loaded with zero

Example
Initially, the value 0xB0 will be loaded in the register R20 and then, a logical shift right will be performed three times. Thus, the following instructions will be executed:

```
LDI R20,0xB0     ; Load the initial value
LSR R20          ; 1st shift
LSR R20          ; 2nd shift
LSR R20          ; 3rd shift
```

Figure 3.41 shows the phases of the multiple shift to the right that is implemented by the above code. The initial value is 0xB0 (176 decimal).
The phases that are shown in figure 3.41 are analyzed as follows:

(a) the initial number is $0xB0=176_{(10)}$
(b) in the first shift, the LSB digit is rejected (is stored in the bit C of the SREG), while the number is filled on the left with a zero
(c) after the first shift, the new content of the register is $0x58=88_{(10)}$
(d) after the second shift, the content of the register is $0x2C=44_{(10)}$
(e) after the third (last) shift the content of the register will be $0x16=22_{(10)}$

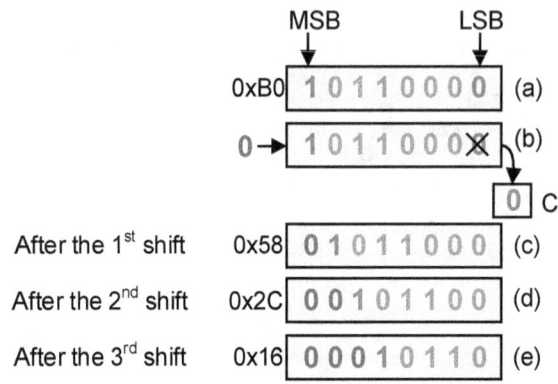

Figure 3.41 Logical shift right

Table 3.10 shows the successive values of the register according to the above shifts.

Table 3.10 Results of the logical shifts (initial number 176)

	Initial number	1st shift	2nd shift	3rd shift
Hexadecimal	B0	58	2C	16
Decimal	176	88	44	22

Based on the table 3.10, in every shift operation the content of the register is divided by 2. The above operation is very important because the division by 2 can be performed with a simple shift instruction.

ROL – Left shift through carry

In the previous shift operations, the MSB or the LSB was rejected while the register was filled with zeros. In a different shift case, the rejected digit fills the register in the next shift. In other words, is a kind of feedback.
The instruction

```
ROL Rd
```

with $d \in [0,31]$

performs a left shift of the register Rd by using the carry. The operation of the above instruction can be described as follows:

Rd(0)=C   Loading the bit 0 from the carry
Rd(n+1)=Rd(n) The bit n is shifting to the next left position
C=Rd(7)   Loading the carry from the bit 7

Example
Figure 3.41 shows how the register content is formed during the successive left shifts through the carry. Initially the value 0x35 will be loaded in the register. The following code tests the ROL instruction operation.

```
LDI R20,0x35       ;Load the initial value
```

```
ROL R20                 ;1st shift
ROL R20                 ;2nd shift
ROL R20                 ;3rd shift
```

Figure 3.42 shows the shift steps during the execution of the above code. Based on the figure 3.42, the code operation is as follows:

(a,b) The initial value is 0x35. Based on the above instruction operation, the carry will become the LSB digit of the register, while the MSB digit will be stored as the new carry.

(c) after the first shift, the new content of the register is 0x6A

(d) in the second shift, the content becomes 0xD4

(e) after the last shift, the content will be 0xA8

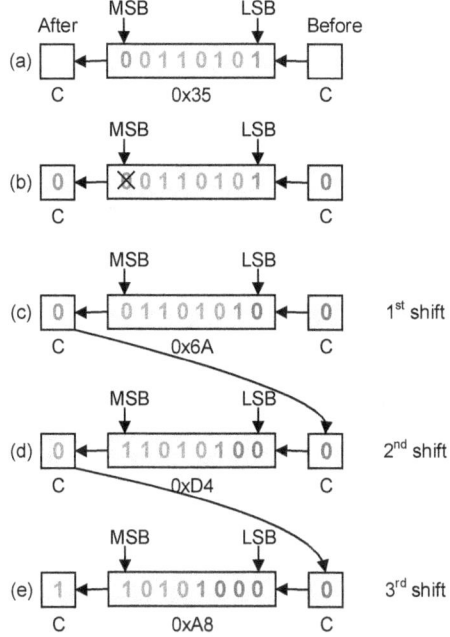

Figure 3.42 Left shift through carry

ROR – Right shift through carry

This shift operation is based on the same logic as the above instruction but the direction is to the right.
The instruction

```
ROR Rd
```

with $d \in [0,31]$

performs a right shift of the register Rd by using the carry. The operation of the above instruction can be described as follows:

Rd(7)=C   Loading the bit 0 from the carry
Rd(n)=Rd(n+1)   The bit n is shifting to the next left position
C=Rd(0)   Loading the carry from the bit 7

ASR – Arithmetic shift right
The instruction
ASR Rd

with d ϵ [0,31]

In this shift operation, the LSB digit is loaded in carry, but only the bits 6 to 0 are involved in the shift operation. Thus, the bit 7 is not affected (fig. 3.43). This is a very useful feature if the MSB represents the sign of the number (if the MSB is involved, then the sign information will be lost).

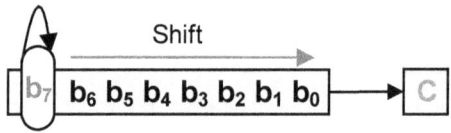

Figure 3.43 Shift with sign prevention

SWAP – Swap High/Low part of a register
The instruction

SWAP Rd

with d ϵ [0,31]

swaps the four most significant bits with the four least significant bits of the register Rd. This operation can be described by writing Rd(7:4)=Rd(3:0) and Rd(3:0)=Rd(7:4).

Example
In this example, the value 0xF0 will be loaded in the register R21 and then, the instruction SWAP will be executed as follows:

```
LDI R21,0xF0      ;Load the initial value
SWAP R21          ;Swap contents
```

The above operation which is described in figure 3.44, is as follows:

(a) the swap operation involves the two 4bit parts (nibble) of the register
(b) the initial number is 0xF0=$11110000_{(2)}$
(c) after swapping, the final number is 0x0F=$00001111_{(2)}$

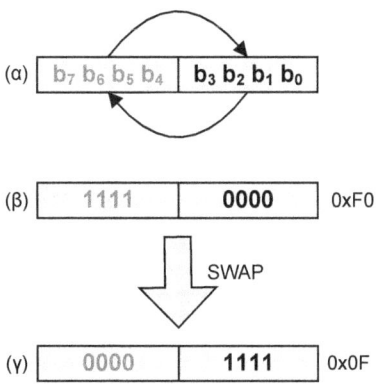

Figure 3.44 Swapping contents

Instructions for manipulating the status register (SREG) contents
In previous section it is mentioned that the status register (SREG) constitutes the most important special purpose register. The SREG consists of independent bits which represent different information like carry, zero result, sign, etc. The SREG structure is presented in figure 3.45 and table 3.11 for one more time.

bit	7	6	5	4	3	2	1	0
information	I	T	H	S	V	N	Z	C

Figure 3.45 SREG structure

Table 3.11 Information bits of the SREG

Bit	Description
I	Activating or deactivating the detection of external interrupts
T	For bit manipulation through the instructions BLD and BST
H	Carry which is produced from the third to fourth bit of a number (e.g. in an ADD instruction)
S	Sign bit. The corresponding value is derived from an exclusive OR between the bits N and V
V	Overflow in a two's complement operation
N	When the MSB digit of a number is 1, this bit is activated (negative result)
Z	When the result of an arithmetic operation is zero, this bit is activated
C	Carry (or borrow in an subtraction) from an arithmetic result

The AVR microcontroller offers many additional instructions for setting (value 1) or clearing (value 0) a selected bit of the status register (SREG).

Figure 3.46 shows the instructions that have access to those bits for setting the corresponding value.

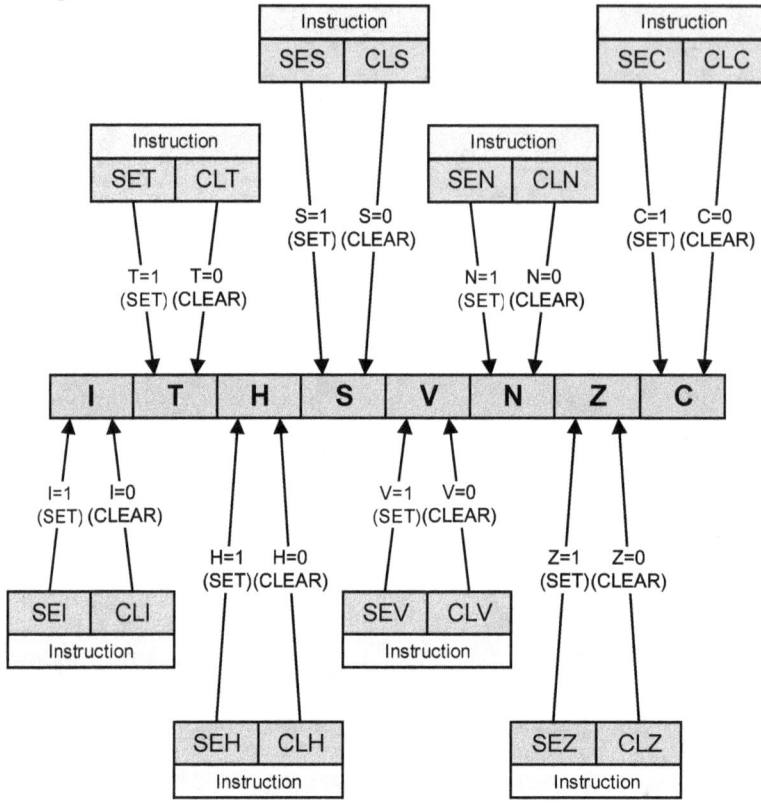

Figure 3.46 Bit manipulation of the SREG

Example

Clearing the bit V. This clearing operation is performed with the instruction CLV (fig. 3.47).

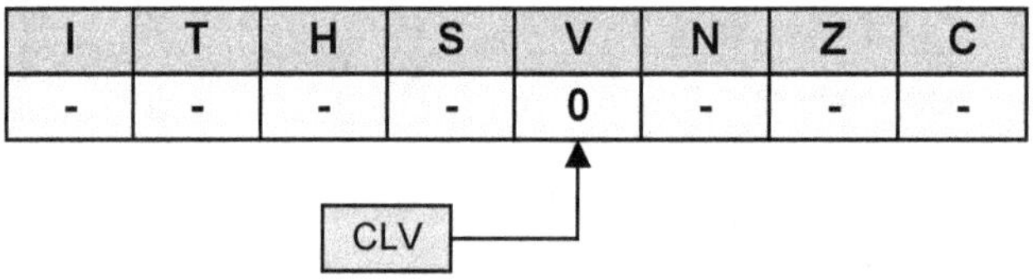

I	T	H	S	V	N	Z	C
-	-	-	-	0	-	-	-

CLV

Figure 3.47 Clearing the bit V

On the other hand, the bit V can be set (value 1) using the instruction SEV (fig. 3.48).

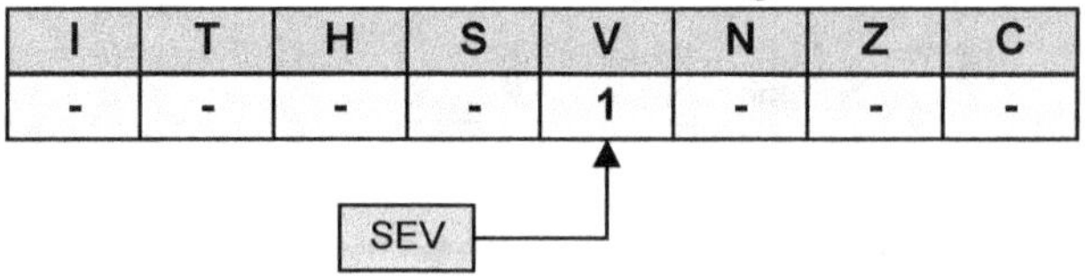

I	T	H	S	V	N	Z	C
-	-	-	-	1	-	-	-

SEV

Figure 3.48 Setting the bit V

Bit manipulation by specifying the position inside SREG

To set or clear a specific bit inside SREG, the instructions `BSET` and `BCLR` can be used.

More precisely, the instruction

`BSET d`
with $d \in [0,7]$

sets (value 1) the bit d of the SREG, while the instruction

`BCLR d`
with $d \in [0,7]$
clears (value 0) the bit d of the SREG.

Example

In this example, the previous instructions will be combined in order to set and clear the bit Z of the SREG successively using the following code:

```
CLZ         ;Clear bit Z (Z=0)
SEZ         ;Set bit Z (Z=1)
CLZ         ;Clear bit Z (Z=0)
BSET 1      ;Set bit Z (Z=1)
BCLR 1      ;Clear bit Z (Z=0)
```

Table 3.12 shows the bit Z manipulation based on the above code. It is obvious that the instruction `CLZ` is equivalent to `BCLR 1`, while `SEZ` is equivalent to `BSET 1`.

Table 3.12 Bit Z manipulation

	bit 7 I	bit 6 T	bit 5 H	bit 4 S	bit 3 V	bit 2 N	bit 1 Z	bit 0 C
CLZ	-	-	-	-	-	-	0	-
SEZ	-	-	-	-	-	-	1	-
CLZ	-	-	-	-	-	-	0	-
BSET 1	-	-	-	-	-	-	1	-
BCLR 1	-	-	-	-	-	-	0	-

Table 3.13 summarizes the shift and bit manipulation instructions.

Table 3.13 Shift and bit manipulation instructions

Instruction	Description	Operation	Bits of the SREG that are affected
LSL Rd	Logical shift left $d \in [0,31]$	C=Rd(7) Rd(n+1)=Rd(n)	Z, C, N, V, H

Instruction	Description	Operation	Flags
LSR Rd	Logical shift right d ∈ [0,31]	Rd(0)=0 C=Rd(0) Rd(n)=Rd(n+1) Rd(7)=0	Z, C, N, V
ROL Rd	Shift left through carry d ∈ [0,31]	Rd(0)=C Rd(n+1)=Rd(n) C=Rd(7)	Z, C, N, V, H
ROR Rd	Shift right through carry d ∈ [0,31]	Rd(7)=C Rd(n)=Rd(n+1) C=Rd(0)	Z, C, N, V
ASR Rd	Arithmetic shift right through carry d ∈ [0,31]	Rd(n)=Rd(n+1), n=0..6	Z, C, N, V
SWAP Rd	Swap High/Low Parts of a register d ∈ [0,31]	Rd(3..0)⇔Rd(7..4)	-
BSET s	Sets a specific bit in the SREG s ∈ [0,7]	SREG(s) = 1	SREG(s)
BCLR s	Clears a specific bit in the SREG s ∈ [0,7]	SREG(s) = 0	SREG(s)
SEC	Sets the bit C of SREG	C=1	C
CLC	Clears the bit C of SREG	C=0	C
SEN	Sets the bit N of SREG	N=1	N
CLN	Clears the bit N of SREG	N=0	N
SEZ	Sets the bit Z of SREG	Z=1	Z
CLZ	Clears the bit Z of SREG	Z=0	Z
SEI	Sets the bit I of SREG	I=1	I
CLI	Clears the bit I of SREG	I=0	I
SES	Sets the bit S of SREG	S=1	S
CLS	Clears the bit S of SREG	S=0	S
SEV	Sets the bit V of SREG	V=1	V
CLV	Clears the bit V of SREG	V=0	V
SET	Sets the bit T of SREG	T=1	T
CLT	Clears the bit T of SREG	T=0	T
SEH	Sets the bit H of SREG	H=1	H
CLH	Clears the bit H of SREG	H=0	H

3.7 Developing complete functional programs

3.7.1 Embedding INC files

In order to develop functional programs and to support the code simplicity, the corresponding code must include some additional directives (instructions to the assembler). For example, for writing in the pins of the port D the number 0xFF, the following instructions will be used:

```
LDI R16,0xFF
OUT PORTD,R16
```

The word PORTD represents actually a physical address which is specific for the AVR model that is used. On the other hand, the programmer does not have to know which is the above physical address. Thus, the corresponding symbolic name is used. For using the above symbolic name, a mapping between the symbolic name and the physical address has to be implemented. For every AVR model, there is a special file (INC file) which defines the corresponding mapping. Table 3.14 shows indicatively, the INC file name for some AVR microcontroller models.

Table 3.14 INC files

Microcontroller	Mapping file
ATmega16	m16def.inc
ATmega32	m32def.inc
ATmega32A	m32Adef.inc
ATmega328	m328def.inc
ATmega8515	m8515def.inc
ATtiny4	tn4def.inc

For example, the mapping of the port D for writing (PORTD) is implemented with the following code:

```
.equ   PORTB = 0x05
```
for the model ATmega328 (inside m328def.inc)

and

```
.equ   PORTB = 0x18
```
for the model ATmega32 (inside m32def.inc)

If the source code is developed within the Atmel Studio software (free software from the ATMEL/Microchip), there is no need to include the mapping file because this happens automatically (when a new project is created, the programmer has to select the corresponding AVR model). On the other hand, if the conversion from the source code to the executable is performed through the command line, then the proper INC file has to be included. For example, in order to include the INC file for the model ATmega32, the following code has to be used:

```
.INCLUDE "include/m32def.inc"
```

Of course, the path to the .inc file must be known.

3.7.2 The basic Assembler directives

The Assembler converts the source code to executable. Except the Assembly instructions that are directly supported by the microcontroller, there is a set of instructions (directives) that are used for helping the programmer during the code development.

Some of the directives that can be used are:

.EQU – Constant value or address definition
With this directive, there is a direct mapping (definition) of a label with a constant value or address. Thus, the above value or address is used through the label (symbolic name).

Example 1 – defining a constant value

```
.EQU COUNTER = 0x01      ;COUNTER=0x01
LDI R16,COUNTER          ;R16=0x01
```

Example 2 – defining an address

```
.EQU COUNTER = 0xFF      ;COUNTER=0xFF
.EQU ADDR = 0x100        ;ADDR=0x100
LDI R16,COUNTER          ;R16=0xFF
STS ADDR,R16             ;[0x100]=0xFF
```

Example 3 – defining different forms of information

```
.EQU VAL1 = 0x10
```
Defining a hexadecimal value

```
.EQU VAL2 = $10
```
Defining a hexadecimal value

```
.EQU VAL3 = 10
```
Defining a decimal value

```
.EQU VAL4 = 0b00001010
```
Defining a binary value

```
.EQU VAL5 = 'A'
```
Defining an ASCII character

.SET – Defining a constant value or address with redefinition capability

The previous directive does not allow to define a new value for an existing label (symbolic name). For having the redefinition capability the directive .SET is used.

Example
```
.SET COUNTER = 0xFA      ;COUNTER=0xFA
LDI R16,COUNTER          ;R16=0xFA
INC R16                  ;R16=R16+1
.SET COUNTER = 0x0A      ;COUNTER=0x0A
LDI R17,COUNTER          ;R17=0x0A
```

.DEF – Assigning a symbolic name to a register

In many cases, the general purpose register names are not helpful for developing the corresponding code by the programmer. With the above directive, symbolic names can be assigned to registers for simplifying the code development.

Example
```
.DEF A=R17    ;A=A new symbolic name for the register R17
.DEF B=R18    ;B=A new symbolic name for the register R18
LDI A,0x44    ;A=0x44
MOV B,A       ;B=A
```

```
INC A           ;A=A+1
```

.ORG – Defining an initial address
The directive .ORG defines the starting address (to be used by the program counter) in which will be stored the first instruction of the code that is follows.

Example
```
.ORG 0            ;Initial address
LDI R16,0xDA      ;First instruction
DEC R16           ;Second instruction
```

.DB – Defining a byte type constant
This directive in combination with a label can be used for defining a byte sequence.

Example
```
NUMDATA .DB 0xFF,0x00,0xAC,0xCC
```

.DW – Defining a word type constant (2 byte)
This directive in combination with a label can be used for defining a double byte (word) sequence.

Example
```
NUMDATA .DW 0x00FF,0x0100,0xB1AC,0x2320
```

EXERCISE SHEETS
Assembly Instructions

1. The values 98$_{(10)}$, 11101010$_{(2)}$ and FC$_{(16)}$ are loading successively to a register. Write the values in the corresponding arithmetic systems.

	16		98	10		2
	16			10	11101010	2
FC	16			10		2

2. Write the needed instructions in order to implement the following operations:

Instruction	Operation
	Load the value A9$_{(16)}$ in the register R21
	Load the value 01011101$_{(2)}$ in the register R25
	Load in R17 the content of R16
	Load in R20 the content of R21

90 CHAPTER 3

3. Assume that R19=0x66 and R20=R19+3. Fill the following gaps:

4. Use the instruction LDS to load in the registers R17,R18,R19 and R20 the contents of the memory locations $0200 to $0203.

5. Use the instruction STS to load in the memory locations $1AF0 to $1AF3 the content of the registers R22, R23, R24 and R29.

6. Write the needed instructions in order to form the following content of the registers:

Content	Instruction or instructions
R17=0xFF	
X=0x0200	

X=0xFABC	
Y=0x1E00	
Y=0x0001	
Z=0xB1E0	
Z=0x45EE	

7. Write the needed instructions in order to implement the following operations:

Load the value 0x0A02, starting from the register R21

Load 2 bytes, starting from the register R21 in the registers pair R1:R0

8. Write the needed instructions in order to load the contents of the memory locations $300 to $305 in the registers R17 to R22 respectively. The address $300 will be written once in the code.

9. The following code is given:
```
LDI R27,03
```

92 ☐ CHAPTER 3

```
LDI R26,0A
LD  R21,X+
LD  R21,X+
LD  R21,X
```

Fill the following gaps during the above code execution.

10. Write the needed instructions in order to load the contents of the memory locations $405 to $400 in the registers R17 to R22 respectively. The address $405 has to be written once in the code.

11. Write the needed instruction in order to load in the register R5 one byte from the memory location $00F0 of the program memory.

12.
Assume that in the program memory location $0100 the number 0x0A0C is stored. Write the needed content of the register Z in order to store in the register R0 the value 0x0C (Low byte of 0x0A0C) by using the LPM instruction.

Z = 0x_____

13.
Assume that in the program memory locations $03AF to $03B5 the numbers 0x00FF, 0x65CC, 0xA000, 0x0000, 0x5321, 0xBCAA and 0xF00A are stored. Write the needed content of the register Z in order to read from the above memory locations (Low or High byte). Fill the following table.

Read High byte	Read Low byte	Z value (Hex)	Z value (Binary)
	0x00FF		
0x65CC			
	0xA000		
0x0000			
	0x5321		
0xBCAA			
	0xF00A		

Write your comments for the binary representations of the register Z

14. The following code is given:

```
LDI    ZL,LOW(digits*2)
LDI    ZH,HIGH(digits*2)
LPM

ADIW   ZL,1
ADIW   ZL,1
ADIW   ZL,1

LPM

digits:  .DB 0x3F,0x06,0x5B,0x4F,0x66,0x6D,0x7D,0x07,0x7F,0x6F
```

94 ☐ CHAPTER 3

Write the content of register R0, during the execution of the above instructions.

R0 = _____ (LPM, first execution)

R0 = _____ (LPM, second execution)

15. Write the needed instructions in order to implement the addition 14+33+22+01+04+08. The result will be stored in the memory location $0091.

16. The following instructions are given:

```
ADD R18,R19
ADD R20,R18
ADD R21,R20
```

Draw a flow chart for describing the addition operation (use the indicative chart for the first addition). It is assumed that R18=1, R19=4, R20=9 and R21=0.

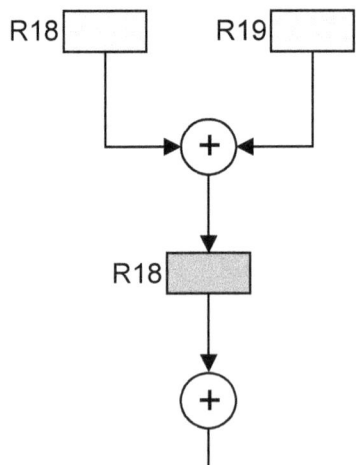

17. Fill the following code with your numbers and write, (a) the final result and (b) the memory location of the final storage

```
LDI  R18,_____

LDI  R19,_____

ADD  R18,R19

LDI  R19,0x00

ADC  R19,R19

STS  $7F,R19

STS  $7E,R18
```

18. Write the needed instructions in order to implement the addition of multiple precision for the numbers:

00000001 10110001₍₂₎, 00000101 10001011₍₂₎

19. Fill the following instructions with your hexadecimal numbers

```
LDI  R30,_____

LDI  R31,_____
```

Write the results that are produced by the following instructions:

Instruction	Operation-Result
ADIW R30,1	

ADIW R30,2	
ADIW R30,3	

20. Load the numbers 0x44, 0xA0, 0xDD, 0x02 and 0x09 in the registers R16, R17, R18, R19 and R20 respectively. Implement the operations AND, OR, XOR, by filling the following tables.

Number	Representation in binary
0x44	
0xA0	
0xDD	
0x02	
0x09	

Operation: R16=R16 AND R17		
Instruction	Result (binary)	Result (hexadecimal)

Operation: R18=R18 OR R20		
Instruction	Result (binary)	Result (hexadecimal)

Operation: R19=R19 XOR R20		
Instruction	Result (binary)	Result (hexadecimal)

21. The following code is given:
```
LDI R19,0xF1
LDI R20,0x77
MUL R19,R20
MOVW R6,R0
```
Which operation is implemented?

Write the result and the storage destination

22. Assume that R20=5. Fill the new contents of the register R20 during the logical instructions execution.

Εντολή	R20 (Hex)	R20 (Binary)
LSL R20		
LSL R20		
LSL R20		
LSL R20		

23. Write the needed instructions for the following calculations:

16/2	R20 + (R23*2)
(R19/2) + (R17*2)	(R17*6) + R19/4
((R16+R17)/2) + R18/2 − R20*4	(R19+R18-R17)/4

** Integer division

24. Develop the needed code for executing the following operations:

(a) Assign the symbolic names REGA, REGB, REGC, REGD and REGD in the registers R18, R19, R20, R21 and R22.

(b) Load the content of the registers REGA, REGB, REGC, REGD and REGE in the registers R0 to R4 respectively

(c) Load in the registers REGA, REGB, REGC, REGD and REGE the numbers 0xFF, 0xDA, 0x00, 0x09 and 0x87 respectively

(d) Load the content of the registers REGA and REGE in the memory locations $387 and $388 respectively

25. Write a program in order to copy the contents of the memory locations $0100-$01F0 in the memory area $0200-$02F0.

26. Modify the previous program in order to set the number of bytes to be copied from the one memory area to the other. The number of bytes is stored in the memory location $0300.

27. Write a program to compute the sum of the contents within the memory area $01E0-$01EF. Store the result in the memory location $0200.

28. Write a program to store in the memory location $02FA the minimum number which is stored in the memory area $01DF-$01FF.

29. Modify the previous program in order to find the maximum number.

30. Write a program to calculate the sums of the memory areas $02A0-$02AF and $03F0-$03FF. After the above calculations, the two sums are compared and in the memory location $0300 the following data are stored:

(a) number 1, if the sums are equal
(b) number 2, if the first sum is less that the second sum
(c) number 3, in any other case

31. Write the needed instructions in order to implement the following circuit logic. The result is stored in the register R18

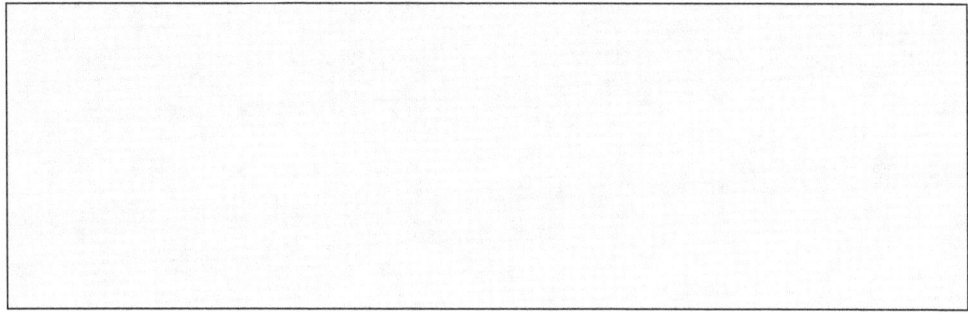

32. Write the needed instructions in order to implement the following circuit logic. The results are stored in the registers R18 and R19.

Note
Some of the previous exercises are for more advanced readers where an additional search within the current book or in the internet is needed.

4 Implementing basic programming structures

Content-Goals
In order to adapt the program to application requirements, the corresponding code must be well structured and organized. The code synthesis is based on the structured programming rules. This chapter presents the basic algorithms and the corresponding code as a tool for developing applications in the right way.

Chapter contents
4.1 Introduction
4.2 Comparison and branch instructions
4.3 Iteration structures
4.4 Absolute and relative jump
4.5 The stack
4.6 Time features of code execution and delay programs
4.7 Macroinstructions
4.8 Program coding in memory

4.1 Introduction

In the programs that have been presented so far, all the instructions are executed sequentially. On the other hand, in a real application there is a need to add "logic" inside the code. This means that some instructions will be executed under specific conditions (e.g. if the content of a register is equal to a numeric value).

Thus, the program development can be implemented using specific rules in an organized framework.

These rules are applied according to the decisions of the programmer for adapting the code to the application specifications and requirements.

As will be shown in the next sections, the development logic (based on well defined steps-algorithm) represents the core of the design. Moreover, the microcontroller instructions implement in practice the above logic.

4.2 Comparison and branch instructions

Based on the structured programming, the program design can be done using three specific structures-rules (sequence, comparison-selection-flow control, iteration).

4.2.1 Basic programming structures
Instruction sequence
Based on that rule, the instructions are executed according to the code sequence (instruction-1 to instruction-n, figure 4.1). This sequence may consist of any number of instructions. Figure 4.1 shows the flow chart diagram as well as the pseudocode.

102 ◻ CHAPTER 4

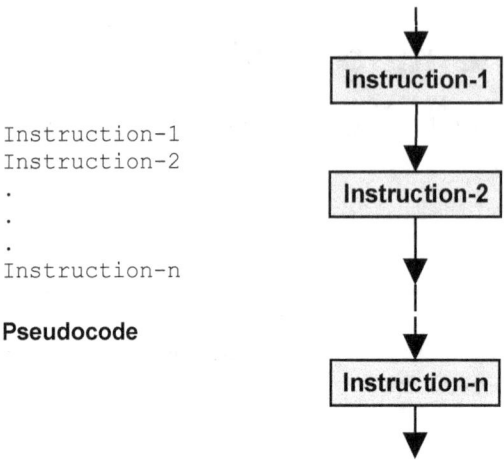

```
Instruction-1
Instruction-2
    .
    .
    .
Instruction-n
```

Pseudocode

Figure 4.1 Instruction sequence

Note
The pseudocode represents the algorithm in a more formal way and describes the final code accurately. This simplifies the development of an operational code, while the algorithmic requirements are met.

Comparison-flow control-selection
The comparison-flow control constitutes a significant code component because supports instruction execution under specific conditions. This programming structure is used in the most of the programs. Figure 4.2 shows the flow chart diagram for the control flow logic as well as the pseudocode.

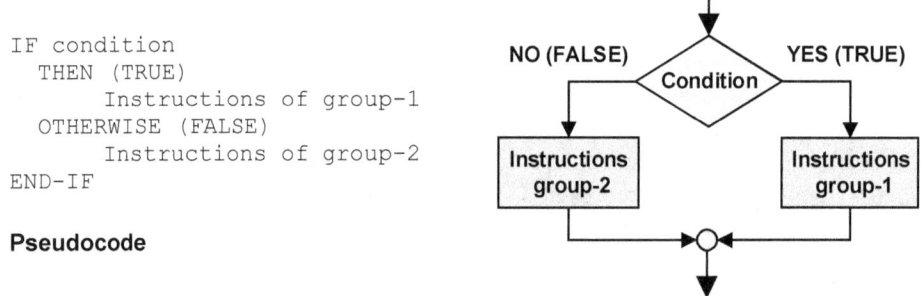

```
IF condition
    THEN (TRUE)
        Instructions of group-1
    OTHERWISE (FALSE)
        Instructions of group-2
END-IF
```

Pseudocode

Figure 4.2 Single control flow

Iteration
The iteration (loop) represents the repetition of an instruction group under one or more conditions. While the iteration conditions are fulfilled, the corresponding instruction group is executed. As shown in figure 4.3, the iteration structure can be implemented using the instruction sequence as well as the control flow structure

(selection). The iteration of figure 4.3, is indicative. As will be shown later there are many alternative implementations of the iteration structure.

```
REPEAT
      Instruction-1
      Instruction-2
      .
      .
      .
      Instruction-n
WHILE the CONDITION is TRUE
```

Pseudocode

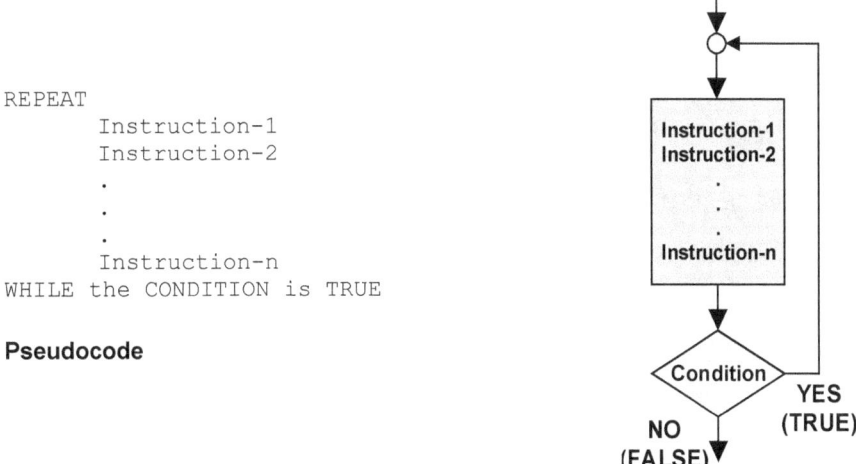

Figure 4.3 Indicative iteration structure

4.2.2 Implementing single control flow structures (selection)

Initially, a single control flow (selection) will be analyzed in order to present how the AVR microcontroller works. Any experienced assembly programmer can easily realize that the implementation philosophy is always the same.

In this example, it is assumed that the code compares the content of a register with an integer value and then, stores a value to a different register under specific conditions. More precisely, the content of register R18 will be compared with the integer value 5. If the register R18 equals to 5, then the operation R19=1 will be performed, otherwise the operation R19=2 (fig. 4.4).

```
IF R18=5
  THEN (TRUE case)
      R19=1
  OTHERWISE (FALSE case)
      R19=2
END-IF
```

Pseudocode

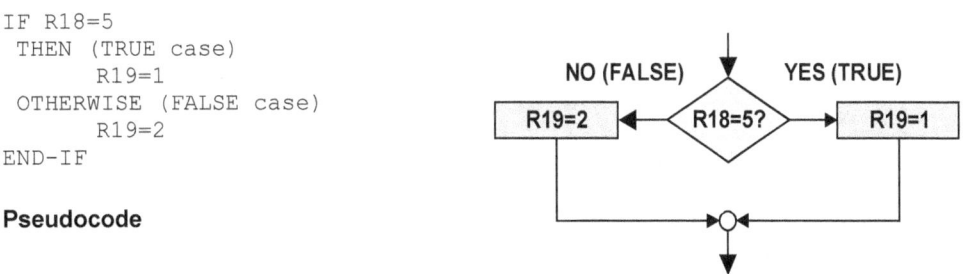

Figure 4.4 An example of control flow (selection)

The following code (code 4.1) implements the above flow control (selection).

Code 4.1

```
CPI R18,5          ;Compare R18 with 5
BREQ iso           ;Branch to label iso if R18=5
LDI R19,2          ;Load the value 2 in R19 (R19=2)
JMP syn            ;Jump to label syn (or RJMP for less bytes in
                   ;program memory) (avoidance of section iso)
iso:               ;iso (label)
        LDI R19,1  ;Load the value 1 to R19 (R19=1)
syn:               ;syn (label)
```

Figure 4.5 shows the possible execution flows.

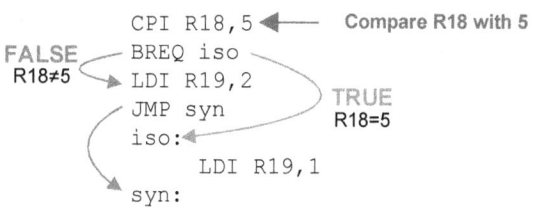

Figure 4.5 Possible execution flows

Based on the above code, initially the comparison of the register R18 with the value 5 (CPI R18,5) is performed. If R18 equals to 5 (EQ-EQual, TRUE case), then a branch is performed to the label iso, and the value 1 is stored in R19 (LDI R19,1). Otherwise, the branch to the iso label is not performed (FALSE case), and the execution flow continues from the next instruction where the value 2 is stored in R19 (LDI R19,2). After this instruction, a jump (JMP) is performed to the syn label in order to avoid the storage of the value 1 in R19 which represents the TRUE case. Figure 4.6 shows the corresponding execution flow for a different initial value of R18. For testing the following code in practice, different initial values have been stored in the register R18.

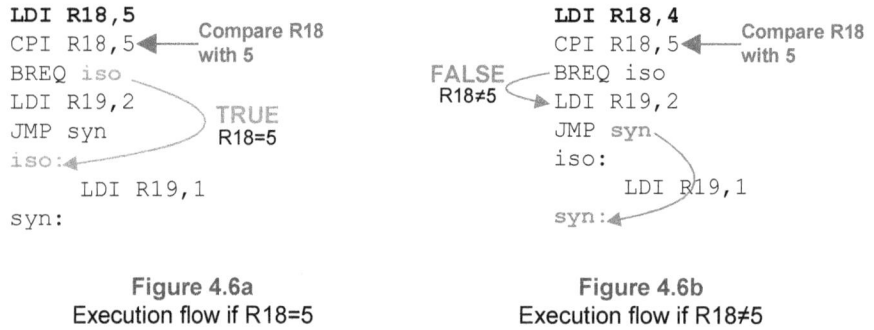

Figure 4.6a
Execution flow if R18=5

Figure 4.6b
Execution flow if R18≠5

Note

For implementing a control flow (selection) programming structure:

(a) A Compare instruction is used for defining what will be checked (e.g. CPI Rd, k for a comparison of a register with an integer value)

(b) The check type is defining by a conditional branch instruction (e.g. BREQ label, for branching to the label, if the corresponding condition is true)
(c) If the condition is false, then the branching to the label label is not performed and the execution flow continues from the next instruction
(d) In special cases, the status register (SREG) bits can be compared directly in order to implement the control flow (selection) with fewer instructions (e.g. in such a case the comparison instruction may be avoided)

The previous example shows the implementation philosophy of the AVR regarding the control flow (selection) programming structure. In the following section, more control flow (selection) instructions will be presented.

CPI – Compare a register with an integer value
The instruction

```
CPI Rd,k
```
with $d \in [16,31]$, $k \in [0,255]$

performs a comparison between the content of the register Rd and the integer value k, without affecting the register. Behind this instruction, the microcontroller subtracts the integer value from the register and activates some bits in the status register (SREG). The execution flow is driven to a specific label (code point) based on a branch condition (e.g. BREQ).

CP – Compare a register with a register
This type of comparison is performed with the instruction

```
CP  Rd,Rs
```
with $d,s \in [0,31]$

The only difference with the previous instruction is the second argument. This instruction is followed by a conditional branch.

CPC – Compare registers with carry
The instruction

```
CPC Rd,Rs
```
with $d,s \in [0,31]$

compares two registers and takes also in account the carry value. Thus the comparison inside the microcontroller is performed by the calculation Rd-Rs-C.

Basic conditional branch instructions
As mentioned previously, after the comparison instruction, a conditional branch is followed. The symbolic name (mnemonic) of a branch instruction starts with the

letters BR (represent the word BRANCH) that are followed by two more letters which represent the type of check.

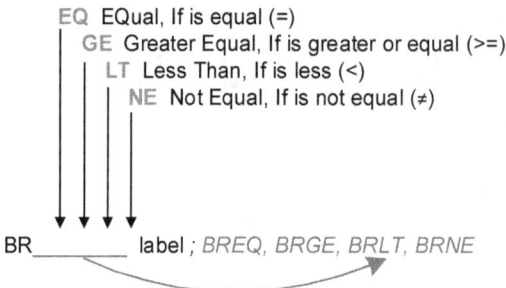

EQ EQual, If is equal (=)
GE Greater Equal, If is greater or equal (>=)
LT Less Than, If is less (<)
NE Not Equal, If is not equal (≠)

BR_____ label ; BREQ, BRGE, BRLT, BRNE

Example 1

It is assumed that the register R17 will be compared with zero, and based on the result, an integer value will be stored in the register R18. Figure 4.7 shows the algorithm (pseudocode) as well as the flow chart diagram.

```
IF R17 = 0, THEN (R17 is zero)
   R18=1
Otherwise
   IF R17 < 0, THEN (R17 is negative)
      R18=2
   Otherwise (R17 is positive)
      R18=3
   END-IF
END-IF
```

Pseudocode

Figure 4.7 Triple check with zero

Figures 4.8a to 4.8c show the possible execution flows based on the R17 content (zero, negative, positive). If the content of register R17 is zero (fig. 4.8a), then the condition of the BREQ is TRUE and the execution flow continues from the label iso. After the point iso, the value 1 is stored in the register R18 (LDI R18,1) and the execution flow is continued from the point cont (JMP cont) in order to bypass the mik section which corresponds to a different case.

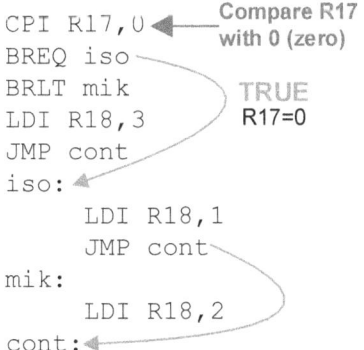

Figure 4.8a Execution flow for R17=0

On the other hand, if the content of register R17 is less than zero (fig. 4.8b), then the condition of the BREQ is FALSE and the execution flow is not driven to the label iso. Thus, the instruction BRLT mik is executed. Now, this condition is TRUE (R17<0) and the execution flow continues from the point mik where the value 2 is stored in R18 (LDI R18,2). After the execution of this instruction, the execution flow continues from the point cont which follows (is a wrong programming practice the insertion of a JMP instruction for going to the cont point because the execution flow is driven always there).

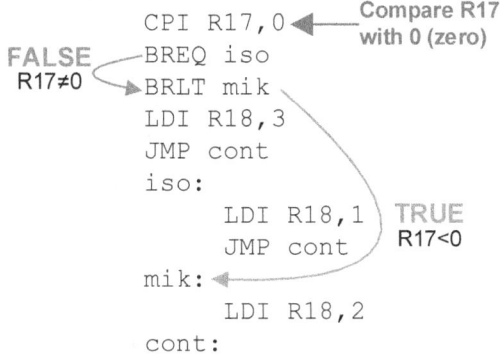

Figure 4.8b Execution flow for R17<0

Finally, if the content of register R17 is positive (R18>0), then the conditions of the instructions BREQ and BRLT are FALSE. Thus, the execution flow is not driven to the points iso or mik and the instruction LDI R18,3 is executed. Initially, the condition for equivalence (BREQ) is checked which is FALSE. Thus, the next condition is checked which is also FALSE. As a result, the value 3 is stored in the register R18. After the value storage, the instruction JMP cont is executed for bypassing the next code sections which belong to different cases (R17=0, R17<0), as shown in figure 4.8c.

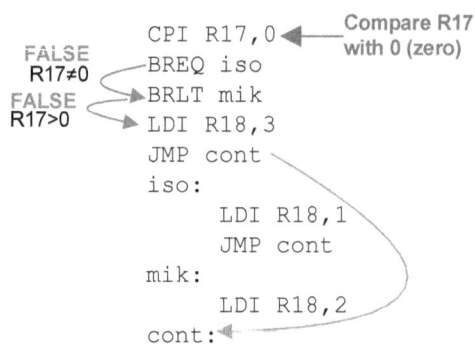

Figure 4.8c Execution flow for R17>0

Note
It must be noticed that despite the three cases, only two conditions are implemented. This happens because if the first two conditions are false, then the third case is definitely true (>0). Thus, less instructions are used. Moreover, such an approach is correct from a programming practice point of view.

If the above instructions are studied in depth based on the microcontroller specifications, the conditional branches can be expressed with an another way:

```
BREQ k              ;branch (set execution flow) +/- k memory locations if is equal
```

with $k \in [-64, 63]$

Despite the fact that for simplicity reasons, labels are used for setting reference points inside code, in practice, a number of steps for changing (+/-) the contents of Program Counter (PC) is set in order to drive the execution flow to the desired memory location. The value of k is added or subtracted from the content of the PC and the new content is PC=PC+k+1 (fig. 4.9). Thus the corresponding branch is called relative.

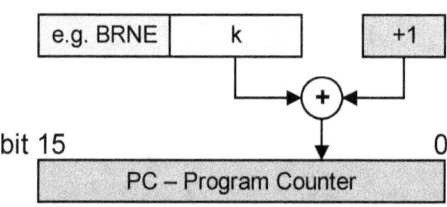

Figure 4.9 Address calculation for the PC

In practice, there is a distance limitation (memory addresses) where the execution flow (through the PC) can be driven.

For better understanding of the above limitation, a code will be developed which generates an error because the label is located out of the permitted address (distance) regarding the instruction execution. In the following code the instruction NOP (No Operation) which just takes one machine cycle without doing any operation and occupies a program memory location is inserted. Thus, using many times the above instruction the mentioned limitation can be studied.

The test code follows (code 4.2):

Code 4.2

```
CP R20,R21    ;Hypothetical comparison
BREQ here     ;Branch to the label here, if are equal
nop           ;1st nop instruction
nop           ;2nd nop instruction
;...
;...
;...
nop           ;62nd nop instruction
nop           ;63rd nop instruction
nop           ;64th nop instruction
here:
     nop
```

The above NOP instructions occupy 64 locations of program memory. The length of the NOP instruction is equal to a program memory location length which is 16bits. With the BREQ here instruction, a branch to the label here which is more than 64 memory locations away is attempted. Thus, an error is generated. Figure 4.10 shows the deviation of the program counter (PC) in order to drive the execution flow to the label here. It is assumed that the instruction BREQ here, is initially executed (the PC points to the corresponding address). Thus, for continuing the execution flow from the label here, the PC must point 65 memory locations away (after the 64 memory locations).

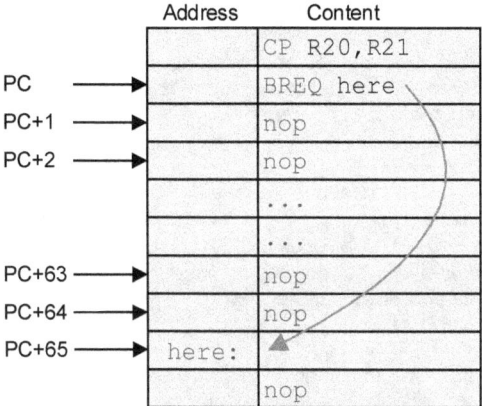

Figure 4.10 Branching out of bounds

The same limitation is also applied for the rest of the conditional branch instructions (BRGE, BRLT, BRNE). Table 4.1 shows the supported branch instructions by the microcontroller AVR. Using this table, many programming structures and applications can be developed.

Table 4.1 Conditional branch instructions

Instruction	Description	Operation	Bits of the SREG which be affected
BREQ k	Branching k locations forward or backward in relation to the program counter (PC). The bit Z of the SREG is checked and if is activated, then the branching operation is performed (equivalence detection, Rd = Rs). The k can be a label. $-64 \leq k \leq +63$	If Rd=Rs, then PC=PC+k+1, otherwise PC=PC+1	-
BRGE k	Branching k locations forward or backward in relation to the program counter (PC). The bit S of the SREG is checked and if is not activated (thus Rd ≥ Rs), then the branching operation is performed. The k can be a label. $-64 \leq k \leq +63$	If Rd≥Rs, then PC=PC+k+1, otherwise PC=PC+1	-
BRHC k	Branching k locations forward or backward in relation to the program counter (PC). The bit H of the SREG is checked and if is not activated, then the branching operation is performed. The k can be a label. $-64 \leq k \leq +63$	If H=0, then PC=PC+k+1, otherwise PC=PC+1	-
BRHS k	Branching k locations forward or backward in relation to the program counter (PC). The bit H of the SREG is checked and if is activated, then the branching operation is performed. The k can be a label. $-64 \leq k \leq +63$	If H=1, then PC=PC+k+1, otherwise PC=PC+1	-
BRID k	Branching k locations forward or backward in relation to the program counter (PC). The bit I of the SREG is checked and if is not activated, then the branching operation is performed. The k can be a label. $-64 \leq k \leq +63$	If I=0, then PC=PC+k+1, otherwise PC=PC+1	-
BRIE k	Branching k locations forward or backward in relation to the program counter (PC). The bit I of the SREG is checked and if is activated, then the branching operation is performed. The k can be a label. $-64 \leq k \leq +63$	If I=1, then PC=PC+k+1, otherwise PC=PC+1	-
BRLO k	Branching k locations forward or backward in relation to the program counter (PC). The bit C of the SREG is checked and if is activated, then the branching operation is performed (Rd < Rs). The k can be a label. $-64 \leq k \leq +63$	If Rd<Rs, then PC=PC+k+1, otherwise PC=PC+1	-

BRLT k	Branching k locations forward or backward in relation to the program counter (PC). The bit S of the SREG is checked and if is activated (thus Rd < Rs), then the branching operation is performed. The k can be a label. $-64 \leq k \leq +63$	If Rd<Rs, then PC=PC+k+1, otherwise PC=PC+1
BRMI k	Branching k locations forward or backward in relation to the program counter (PC). The bit N of the SREG is checked and if is activated, then the branching operation is performed. The k can be a label. $-64 \leq k \leq +63$	If N=1, then PC=PC+k+1, otherwise PC=PC+1
BRNE k	Branching k locations forward or backward in relation to the program counter (PC). The bit Z of the SREG is checked and if is not activated, then the branching operation is performed (Rd <> Rs). The k can be a label. $-64 \leq k \leq +63$	If Rd<>Rs, then PC=PC+k+1, otherwise PC=PC+1
BRPL	Branching k locations forward or backward in relation to the program counter (PC). The bit N of the SREG is checked and if is not activated, then the branching operation is performed. The k can be a label. $-64 \leq k \leq +63$	If N=0, then PC=PC+k+1, otherwise PC=PC+1
BRSH k	Branching k locations forward or backward in relation to the program counter (PC). The bit C of the SREG is checked and if is not activated, then the branching operation is performed. The k can be a label. $-64 \leq k \leq +63$	If Rd≥Rs, then PC=PC+k+1, otherwise PC=PC+1
BRTC k	Branching k locations forward or backward in relation to the program counter (PC). The bit T of the SREG is checked and if is not activated, then the branching operation is performed. The k can be a label. $-64 \leq k \leq +63$	If T=0, then PC=PC+k+1, otherwise PC=PC+1
BRTS k	Branching k locations forward or backward in relation to the program counter (PC). The bit T is checked and if is activated, then the branching operation is performed. The k can be a label. $-64 \leq k \leq +63$	If T=1, then PC=PC+k+1, otherwise PC=PC+1
BRVC k	Branching k locations forward or backward in relation to the program counter (PC). The bit V of the SREG is checked and if is not activated, then the branching operation is performed. The k can be a label. $-64 \leq k \leq +63$	If V=0, then PC=PC+k+1, otherwise PC=PC+1

BRVS k	Branching k locations forward or backward in relation to the program counter (PC). The bit V of the SREG is checked and if is activated, the branching operation is performed. The k can be a label. $-64 \leq k \leq +63$	If V=1, then PC=PC+k+1, otherwise PC=PC+1

Sorting the branch instructions based on the SREG bit (table 4.1), the table 4.2 is produced.

Table 4.2 Bit check

Instruction	Bit value of the SREG for branching to another location
BRNE	Z=0
BREQ	Z=1
BRPL	N=0
BRMI	N=1
BRSH	C=0
BRLO	C=1
BRVC	V=0
BRVS	V=1

4.3 Iteration structures

4.3.1 Iteration structure do-while

In this iteration type, the condition check is performed at the end. As a result, the instruction group inside the main section of the iteration is executed at least once. The iteration is continued as the conditional expression at the end of the main section is true. For better understanding of the corresponding implementation, the above iteration type will be developed both in C and assembly language (fig. 4.11).

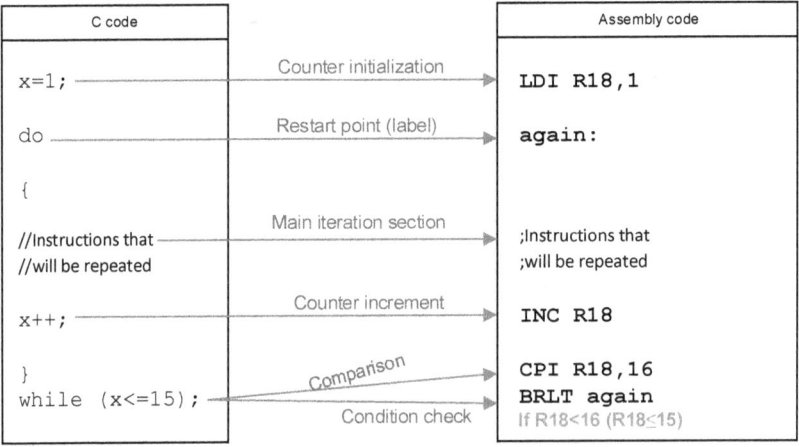

Figure 4.11 C and Assembly code

As shown in figure 4.11, the iteration structure implementation always consists of the same components independently of the programming language. The counter initialization is performed before the main iteration section (restart point) in order not to produce an endless loop. Inside the iteration section, instructions that will be repeated are placed (are executed at least once) while the loop termination is implemented with the counter update and the conditional check (if the condition is TRUE, then the execution flow returns to the label do for the C code or the label again for the assembly code). The counter takes values in the range [1,15] (counter>15 after the loop). Using the instruction BRLT (less than) and check value 16, the *if condition R18≤15* is implemented (the allowed values are <16). Thus, if this condition is TRUE, then the iteration is continued. Figure 4.12 shows the flow chart diagram of the code operation regarding the mentioned programming structure.

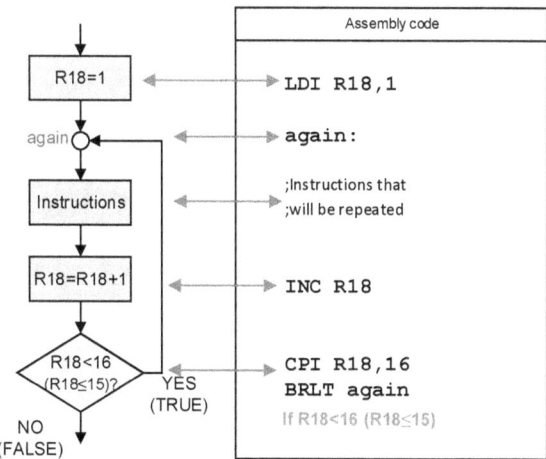

Figure 4.12 Code and flow chart diagram

In previous implementations, the comparison was performed with an integer value which was greater than the initial value. If the above specific values of the counter are not critical (e.g. are not used in a calculation) but only the number of iterations, then the above comparison can be avoided. In such a case the comparison with zero can be performed through the instruction BRNE which checks the Z bit (Zero flag) of the SREG (while the counter value is not zero). It must be noticed that for implementing the above logic, the starting value of the counter has to be greater than zero and the instruction INC to be replaced by a DEC instruction.

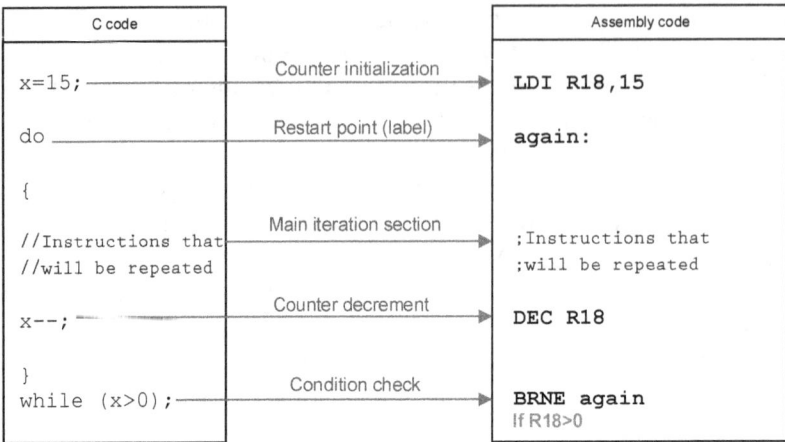

Figure 4.13 C and Assembly

As shown in figure 4.13, the counter takes values in the range [15,0], while the loop will be terminated when the counter content becomes zero (the Z bit of the SREG will be activated). Figure 4.14 shows the corresponding flow chart diagram.

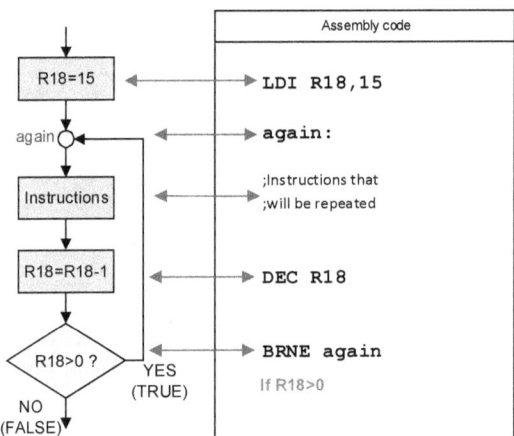

Figure 4.14 Code and flow chart diagram

4.3.2 Iteration structure while-do

In this type of iteration, the condition check is performed at the beginning (before the main iteration section). As a result, the instruction group inside the iteration, is never executed if the condition is not TRUE at least once. The implementation of the above structure is more complex as compared to the previous (do-while). Moreover, the while-do structure can be implemented with two different ways.

Figure 4.15 shows the flow chart diagram, while the code 4.3 is the implementation of the iteration structure where the execution flow is driven in the main iteration section only if the condition check is TRUE. Otherwise, the iteration is terminating with jumping to the label exit, where the condition check is performed. When the condition check is FALSE, the execution flow is not driven to the label start (main iteration section) and thus the instruction JMP exit is performed for exiting from the iteration structure.

On the other hand, in the second implementation (fig. 4.16, code 4.4), is checked if the counter value is greater than 15 (≥16). In such a case, the execution flow is driven to the label exit and the iteration structure is bypassed. Otherwise (FALSE case), the iteration starts due to the fact that the counter value is not greater than 15 (is less or equal to 15).

```
LDI R18,1
again:
        CPI R18,16
        BRLT start
        JMP exit
start:
        ;Instructions
        ;that will be
        ;repeated
        INC R18
        JMP again
exit:
```

Code 4.3

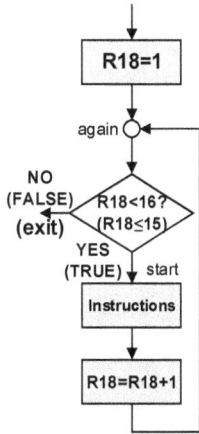

Figure 4.15 Iteration with condition TRUE

```
LDI R18,1
again:
        CPI R18,16
        BRGE exit
        ;Instructions that
        ;will be repeated
        INC R18
        JMP again
exit:
```

Code 4.4

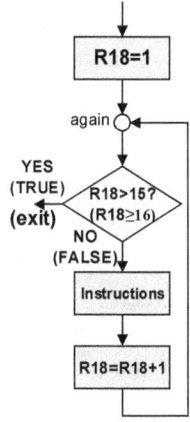

Figure 4.16 Iteration with condition FALSE

4.3.3 Nested loop (iteration structure)

In many cases, there is a need for a large number of iterations in order to implement a time delay, numbers manipulation, etc. The fact that 8bit registers are used, limits the number of iterations. For unsigned 8bit numbers the corresponding range is 0 to 0xFF (0 to 255). Thus, for more iterations, two or more iteration structures (loops) have to be used. Using a loop into another (nested loop), the number of iterations is the product of iterations of the two loops. For 25 iterations, 5 iterations will be needed in each loop. For implementing two loops of 5 iterations, the corresponding structure is as follows:

```
For A=1 to 5
        For B=1 to 5
                Instructions that will be repeated
                25 times in total (5*5)
        End-B
End-A
```

The corresponding pseudocode of the above structure is as follows:

Pseudocode
```
A=1
Repeat
   B=1
   Repeat
        Instruction group-A
        B=B+1
   WHILE B<6  (≤5)
   Instruction group-B
   A=A+1
WHILE A<6  (≤5)
```

The above pseudocode consists of all the necessary components in order to describe the corresponding algorithm which is critical for the code development. As shown in the above pseudocode, there are two counters, one for each loop. Each counter is initialized before the corresponding loop starting point and is checked before the end of the iteration structure.

Figure 4.17 shows the flow chart diagram as well as the code. The points 3 to 7 represent the nested loop (interior loop), while the points 1,2 and 8 to 10 represent the outer loop. For each execution cycle of the outer loop, 5 executions cycles of the nested loop are performed. The counter initialization regarding the outer and the nested loop is performed at the points 1 and 3 respectively. The `instruction group-A` will be executed 25 times in total due to the fact that belongs to the nested loop. On the other hand, the `instruction group-B` will be executed 5 times in total due to the fact that belongs to the outer loop.

Implementing basic programming structures 117

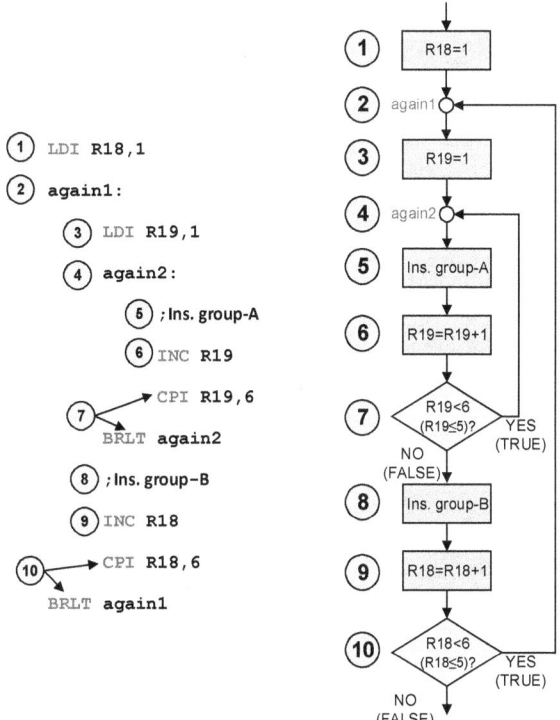

```
① LDI R18,1
② again1:
    ③ LDI R19,1
    ④ again2:
        ⑤ ;Ins. group-A
        ⑥ INC R19
        ⑦ CPI R19,6
           BRLT again2
    ⑧ ;Ins. group-B
    ⑨ INC R18
    ⑩ CPI R18,6
       BRLT again1
```

Figure 4.17 Code and flow chart diagram

The final code of the above structure is as follows (code 4.5):

Code 4.5

```
LDI R18,1           ;Counter initialization (outer loop)
again1:             ;Restart point of the outer loop
   LDI R19,1        ;Counter initialization (inner-nested loop)
   again2:          ;Restart point of the inner-nested loop
      ;Ins. group-A;Instructions that will be repeated 25 times
      INC R19       ;Counter increment (inner-nested loop)
   CPI R19,6        ;Compare with the check value
   BRLT again2      ;Continue the iteration if the
                    ;limit value is not reached
   ;Ins. group-B    ;Instructions that will be repeated 5 times
   INC R18          ;Counter increment (outer loop)
CPI R18,6           ;Compare with the check value
BRLT again1         ;Continue the iteration if the
                    ;limit value is not reached
```

4.3.4 Implementing a loop without comparison

The AVR microcontroller supports a large number of branching instructions which check specific bits of the SREG. Using directly some SREG bits, a loop can be developed without a comparison instruction. For example, any arithmetic operation (e.g. DEC R18) affects specific bits of the SREG. If the instruction DEC R18 (R18=R18-1) is executed, then the bit Z of SREG can be activated (when R18=0). On the other hand, the instruction BRNE checks the bit Z of the SREG and if is not activated, then the branching operation is performed to the label which is the instruction argument. The following codes (codes 4.6a,b), performs the same number of iterations but with different implementation.

```
LDI R18,1
again1:
        LDI R19,1
        again2:
                ;Ins. group-A
                INC R19
                CPI R19,3
        BRLT again2
        ;Ins. group-B
        INC R18
        CPI R18,3
BRLT again1
```

Code 4.6a
Implementation by checking a positive value

```
LDI R18,2
again1a:
        LDI R19,2
        again2a:
                ;Ins. group-A
                DEC R19
        BRNE again2a
        ;Ins. group-B
        DEC R18
BRNE again1a
```

Code 4.6b
Implementation by checking a zero case

In the code 4.6a, the instruction BRLT checks the S bit (Sign flag) of the SREG. When the instruction CPI R18,3 is executed, the microcontroller performs the operation R18-3. If R18<3, then the subtraction result is negative. Thus, S=1 and the branching is performed to the label again1. The same logic is used also for the execution of the instruction CPI R19,3. On the other hand, in the code 4.6b, before the branch instruction (BRNE) there is an arithmetic instruction for decreasing the counter content by 1. The instruction BRNE checks the bit Z of the SREG which is not active while the previous instruction does not give a zero result (R18 is not equal to zero). The counter content is reaching the zero, and thus will be zero after some iterations. Using a branch instruction which is based directly on the bit Z of the SREG, a comparison instruction is avoided. Figure 4.18 shows the algorithmic representation (flow chart diagram) for the implementation of the two loops (one nested loop) by checking directly the Z bit.

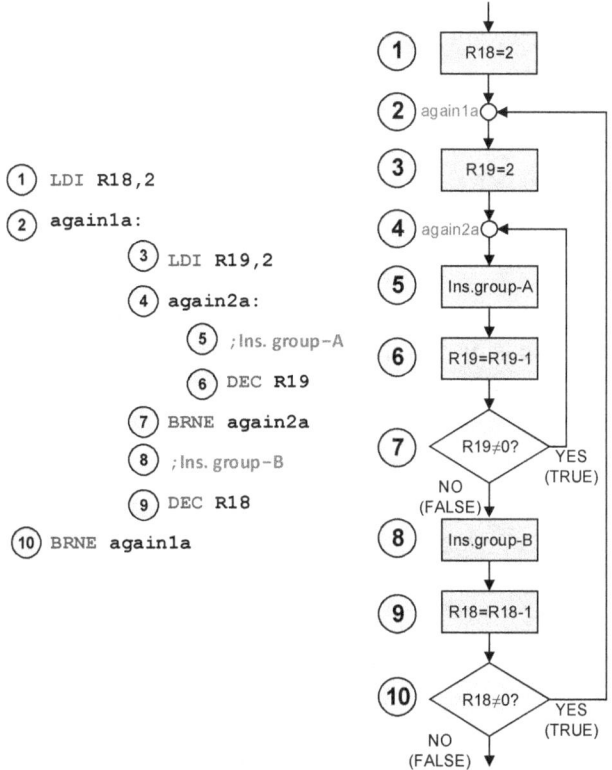

```
①  LDI R18,2
②  again1a:
③          LDI R19,2
④          again2a:
⑤                  ;Ins. group-A
⑥                  DEC R19
⑦                  BRNE again2a
⑧          ;Ins. group-B
⑨          DEC R18
⑩  BRNE again1a
```

Figure 4.18 Two loops (checking the Z bit)

4.3.5 Defining the number of iterations beyond the register limit

The method for calculating the total number of iterations has been presented in the previous section. Which is the maximum number of iterations by using two loops? A loop counter is implemented by a general purpose register (e.g. R16) and thus the maximum number of iterations for a loop is 256 ([0x00-0xFF], 8bit). Using a double loop (one nested loop) with two counters, the number of iterations in total is $256*256=256^2=65536$ (0xFFFF, 16bit). In order to increase the number of iterations (beyond 16bit) more loops have to be implemented (with more counters).

<u>Example</u>
It is assumed that there is a need for 350.000 iterations in total. For counting to that number, 19bits ($N=\log_2 350.000$) are needed (rounding the N to the upper integer). Thus, three 8bit registers will be used (3x8bit=24bit). Using all the possible range of three registers, the total number of iterations will be $256^3=16777216$. For achieving the needed number of iterations (350.000) the proper register ranges have to be chosen.

120 CHAPTER 4

In the following structure, a specific range for every register is chosen in order to achieve the needed number of iterations.

```
For A=1 to 35
    For B=1 to 100
        For C=1 to 100
            Instructions that will be repeated
            350.000 in total (35*100*100)
        END-C
    END-B
END-A
```

Before the final assembly code implementation, a more representative pseudocode will be developed for focusing mainly on the counter decrement. The code 4.7 is the implementation of the pseudocode, while the figure 4.19 shows the flow chart diagram.

Pseudocode
```
A=35
Repeat
    B=100
    Repeat
        C=100
        Repeat
            Ins. group-A
            C=C-1
        WHILE C≠0
        B=B-1
    WHILE B≠0
    A=A-1
WHILE A≠0
```

Code 4.7
```
LDI R16,35
start1:
    LDI R17,100
    start2:
        LDI R18,100
        start3:
            ;Ins. group-A
            DEC R18
            BRNE start3
        DEC R17
    BRNE start2
    DEC R16
BRNE start1
```

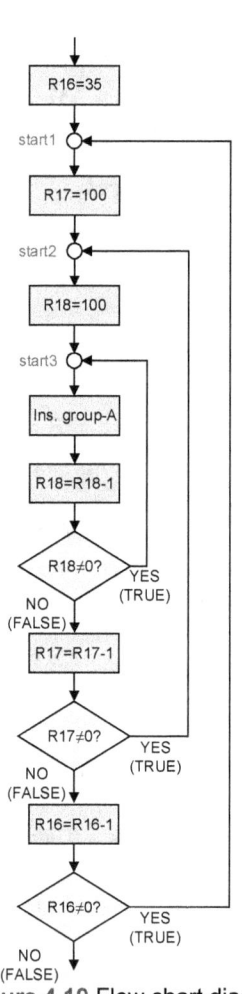

Figure 4.19 Flow chart diagram

4.4 Absolute and relative jump

As shown in the previous examples, the instruction JMP is used for jumping without any condition (unconditional jump) to a specific code point (label). For better understanding, a previous example is presented once more.

Code 4.8

```
CPI R18,5          ;Compare R18 with 5
BREQ iso           ;If R18=5, then branch to iso
LDI R19,2          ;otherwise, R19=2
JMP syn            ;Jump to syn (bypass the LDI R19,1)
iso:
      LDI R19,1    ;R19=1
syn:
```

The same example can be developed by replacing the instruction JMP with the instruction RJMP.

Code 4.9

```
CPI R18,5          ;Compare R18 with 5
BREQ iso           ;If R18=5, then branch to iso
LDI R19,2          ;otherwise, R19=2
RJMP syn           ;Jump to syn (bypass the LDI R19,1)
iso:
      LDI R19,1    ;R19=1
syn:
```

Which is the difference between JMP and RJMP;
The differences between the above instructions can be summarized as follows:

(a) The JMP performs a jump (absolute) in a specific address in the whole memory, while RJMP increases or decreases the content of the program counter (PC) by k which belongs to a specific range (relative jump) and limits the distance (memory locations) of the jump.

(b) The argument of JMP may be any address inside the program memory. A JMP instruction occupies two memory locations and three machine cycles are needed for the execution. On the other hand, RJMP occupies only one memory location and only two machine cycles are needed for the execution.

For the above reasons, the RJMP instruction is usually used. In the following section, more details for the jump instructions will be presented.

The JMP instruction is expresses with one argument as follows:

JMP addr

The addr is an address which is expresses with 22bit. When this instruction is executed, the operation PC=addr is performed (the execution flow is driven to the address `addr`). Using 22bit for the address, the `JMP` instruction can be refer to any memory location within the program memory.

In most of the 8bit AVR models, the program memory is ranged in some hundreds of Kbytes. Thus, a `JMP` instruction with a 22bits argument is excessive. Moreover, the target locations of a jumping instruction are usually close to the jump instruction. The `RJMP` instruction is expresses as follows:

`RJMP k`

The `k` is ranged in 12bit. Due to the fact that the jump will be performed to previous (lower) or next (higher) memory addresses, the `k` must be a signed number. Thus, only 11 bits ($2^{11}=2048$) will be used for the address number. With this approach, the address steps will be such, in order the `k` to belong in the range [-2048, 2047].

When `k` is a label, is automatically replaced with a real memory address (absolute or relative jump).

Indirect jump

Despite the differences of the above jump instructions, there is a common limitation between them. This limitation is that the argument (address) is always constant in the program. On the other hand, in some cases there is a need for adapting the jump address to current program execution status. The above adaptation means that the corresponding argument must be variable. For the above reasons, the instruction `IJMP` which is expressed without arguments is used. The address of jump is read from the register Z which is 16bits and is formed from the registers pair R31:R30. Thus, the jump address can be changed by changing the content of the register Z. Due to the fact that the program counter (PC) is 22bits long, the content of the register Z will occupy the 16 least significant bits (fig. 4.20).

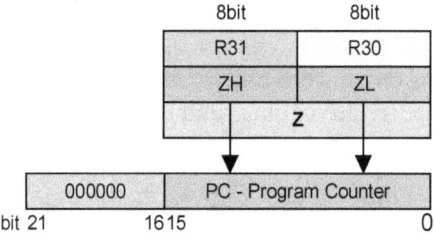

Figure 4.20 Defining the jump address through the register Z

Calling a subroutine in an absolute address
The instruction

`CALL addr`

with *addr* 16 or 22bit (the *addr* can be a label)

The above instruction occupies 32bit in the program memory. The expression of this instruction is similar to the `JMP` instruction. The execution flow returns from the subroutine with the instruction `RET`.

Calling a subroutine in a relative address
If the subroutine is not far away from the current address, then the following instruction can be used:

RCALL k
with k ∈ [-2048,2047] (the *addr* can be a label)

when the above instruction is used, the microcontroller performs the operation PC=PC+k+1.

Indirect call of a subroutine
The instruction

ICALL

reads the address of the subroutine from the content of the register Z (16bit).

Extended indirect call of a subroutine
With the instruction

EICALL

an extended subroutine call is performed where 22bits are used for the address. The least significant 16bits are read from the register Z and the rest of the bits from the register EIND which is not a general purpose register (fig. 4.21). The above instruction has not arguments. Before the instruction execution, the involving registers for defining the target address have to be initialized properly.

Example
```
LDI  R20,0x03
LDI  R31,0x01
LDI  R30,0x02
OUT  EIND,R20
EICALL
```

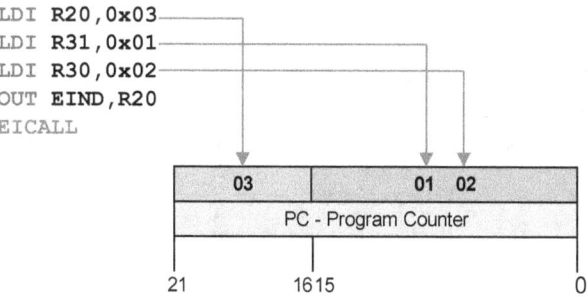

Figure 4.21 Preparing the instruction EICALL

Based on the above example, the target address (where the call will be performed) is 0x030102. Figure 4.22 shows the difference of the instructions ICALL and EICALL regarding the address forming inside the program counter for performing the call.

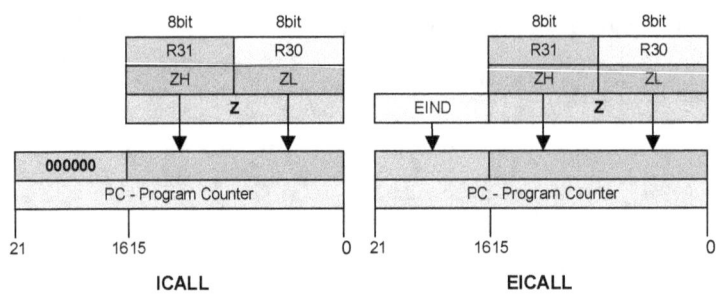

Figure 4.22 ICALL and EICALL

Note
The instruction EICALL is supported by AVR microcontrollers which have 64K or more memory locations.

4.5 The stack

The stack represents a method for organizing data in the memory though the proper software. This organization simplifies the data storage for the programmer as well as the operation of storing addresses in order to call and return from subroutines.

The access of the stack is always performed though the instructions PUSH (store at the top of the stack) or POP (read from the top of the stack). Thus, anything that is stored last, will be read first (LIFO-Last In First Out).

The instruction

PUSH Rs

with Rs ϵ [0,31]

stores at the top of the stack the content of the register Rs, while the instruction

POP Rd

with Rd ϵ [0,31]

reads from the top of the stack and stores the corresponding data in the register Rd.

Moreover, the microcontroller stores the top address of the stack (access point) in the Stack Pointer (SP) register which is formed by two 8bits registers (figure 4.23).

8bit	8bit
SPH	SPL
SP	

Figure 4.23 Stack Pointer (SP)

For a memory of 256bytes long, 8bits are enough for the SP register. Figure 4.24 shows how the register contents (data) are pushed (store) or popped (read) from the stack. Initially, the register contents are R17=0xFA, R18=0x09, R19=0xCB, while the stack is empty. With the first PUSH instruction (PUSH R17), the content of R17

(0xFA) is pushed (store) at the top of the stack (where the SP points) and after this operation the SP content is decreased by 1 in order to point again in an empty location at the top of the stack. The same procedure is also performed for the R18 content (PUSH R18). Thus, the value 0x09 is pushed (store) at the top of the stack and the SP points to the new empty location.

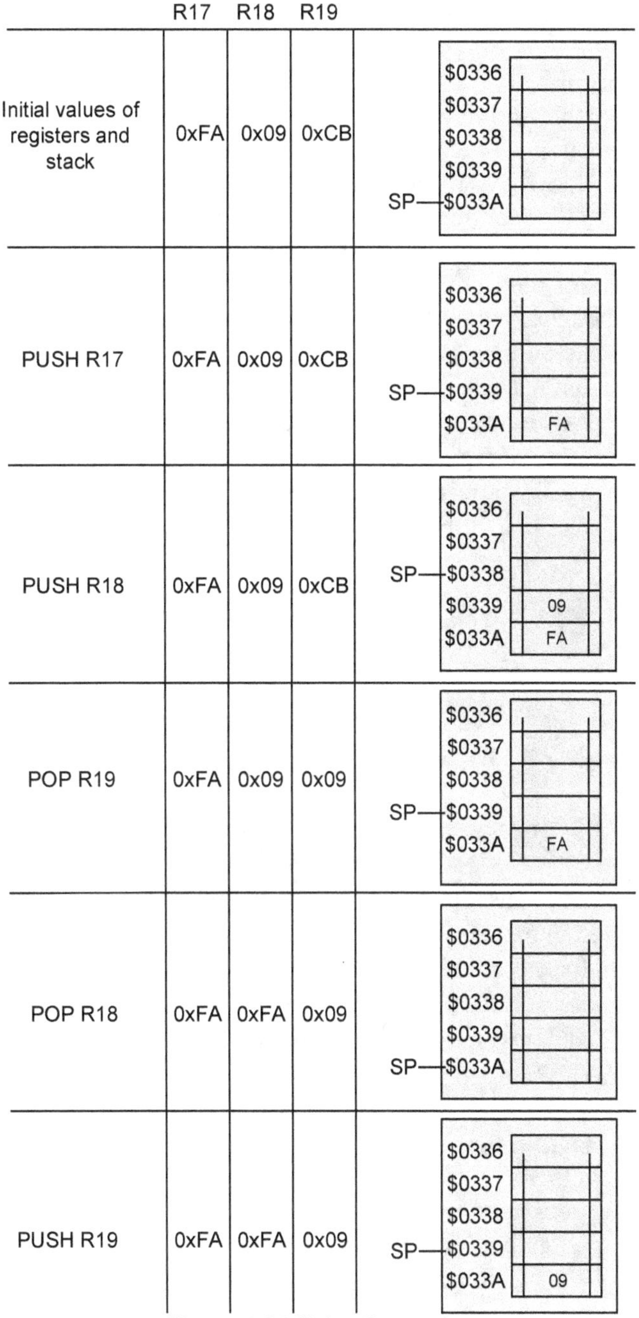

	R17	R18	R19	Stack
Initial values of registers and stack	0xFA	0x09	0xCB	$0336 / $0337 / $0338 / $0339 / SP→$033A
PUSH R17	0xFA	0x09	0xCB	$0336 / $0337 / $0338 / SP→$0339 / $033A = FA
PUSH R18	0xFA	0x09	0xCB	$0336 / $0337 / SP→$0338 / $0339 = 09 / $033A = FA
POP R19	0xFA	0x09	0x09	$0336 / $0337 / $0338 / SP→$0339 / $033A = FA
POP R18	0xFA	0xFA	0x09	$0336 / $0337 / $0338 / $0339 / SP→$033A
PUSH R19	0xFA	0xFA	0x09	$0336 / $0337 / $0338 / SP→$0339 / $033A = 09

Figure 4.24 Using the stack

The next instruction (POP R19) reads from the top of the stack and the value 0x09 is stored in the register R19 (after that, the SP is increased). With another POP instruction the value 0xFA is read from the top of the stack and is stored in the register R18 (after that, the SP is increased). Finally, the PUSH R19 instruction pushes (store) the content of R19 at the top of the stack, while the SP is updated. The PUSH instructions do not change the contents of the corresponding argument.

Stack initialization
The stack is implemented in data memory and has to start from a specific memory location with direction to the low memory addresses (when something is stored in the stack, the SP is decreased). If the stack will be used in a program, then it has to be initialized. Due to the fact that the stack access is achieved through the SP register, the SP has to be initialized correctly. To separate correctly the stack from the main data area (avoiding data corruption), the starting point has to belong to higher addresses. On the other hand, the memory size is different among the AVR models and thus, the higher address is unknown. For solving this issue, the microcontroller has a constant value with the name RAMEND which represents the higher address of the data memory. Moreover, two registers are used for expressing that address. As mentioned before, the SP register is formed by the pair of the registers SPH and SPL. Thus, the stack initialization is based on the constant RAMEND for loading the proper value in the registers pair SPH:SPL. Thus, initialization will be performed with the following code:

Code 4.10

```
LDI R16,HIGH(RAMEND)
OUT SPH,R16
LDI R16,LOW(RAMEND)
OUT SPL,R16
```

The HIGH(RAMEND) represents the high byte (most significant byte) of the higher address, while the LOW(RAMEND) represents the low byte (least significant byte). Using the LDI instructions, the above bytes are stored temporarily in the register R16 and then are stored in the registers SPH and SPL respectively. Thus, the registers pair SPH:SPL is initialized correctly in order to achieve a stack normal operation. It must be noticed that the access to the registers SPH and SPL is performed with the instruction OUT and not with MOV due to the fact that they don't belong to the general purpose registers (R0 to R31).

The role of the stack in the subroutine calls
In many programs there is a need to execute code sections many times. Obviously, the repeated code does not written again and again. Thus, the necessary code section will be written once and will be called whenever is needed. For better understanding the logic behind the subroutine call, the following scenario has been developed (figure 4.25):

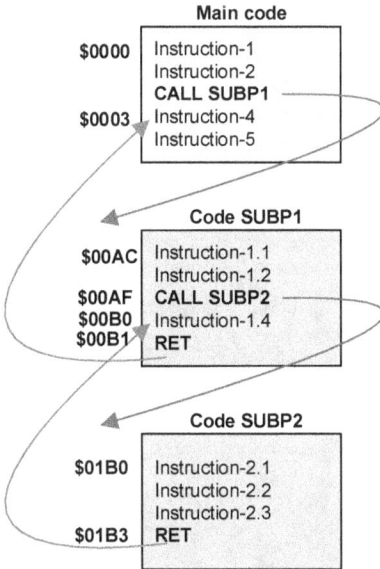

Figure 4.25 Calling subroutines

Figure 4.25 shows the main code as well as two subroutines (SUBP1 and SUBP2) which are executed when they are called. More precisely, after the program start (starting address $0000), the execution flow reaches the instruction "CALL SUBP1". Thus, the execution flow is transferred (through the program counter) to the first instruction of the subroutine SUBP1 (starting address $00AC). Through this subroutine, the SUBP2 is called (starting address $01B0). When RET instruction inside the SUBP2 (address $01B3) is executed, a return is performed to the next instruction of the "CALL SUBP2» (address $00B0). After that, the next RET instruction which belongs to SUBP1 (address $00B1) is executed and the final return to the main code is performed.

From the above example, it is important to investigate the subroutine call methods as well as the stack operation for supporting the return procedure from the subroutines. For understanding the stack operation, a real program based on the figure 4.25 will be developed. Figure 4.26 shows a program which consists of three code parts. The main code starts at the address $0000, the subroutine SUBP1 from the address $00AC and the SUBP2 from the address $01B0. The starting address of the code sections is defined by the directive .ORG. The stack starts from the address $08FF which is the higher hypothetical address of the corresponding microcontroller.

128 CHAPTER 4

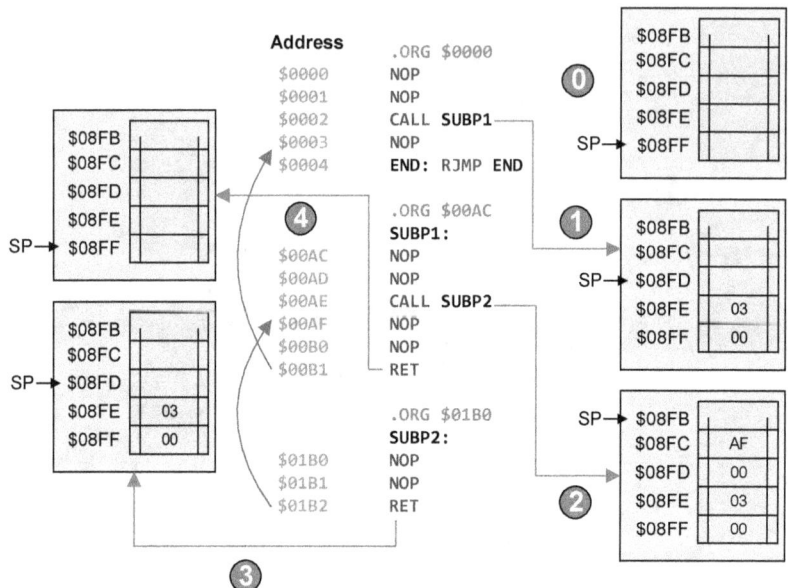

Figure 4.26 Stack operation with subroutine calls

The program operation using the stack, is analyzed as follows:

(0) The program execution starts and the stack is empty

(1) The subroutine SUBP1 is called (address $0002) and the return address ($0003) is pushed (store) in the stack. Thus, the SP register points to the address $08FD.

(2) Inside SUBP1 the subroutine SUBP2 is called (address $00AE) and the return address ($00AF) is pushed (store) in the stack. Thus, the SP register points to the address $08FB.

(3) The RET instruction of the SUBP2 is executed (address $01B2) and the return address ($00AF) is read from the stack (POP). Thus, the execution flow continues from the address $00AF.

(4) After the last RET execution (inside SUBP1, address $00B1), the return address ($0003) is read (POP) from the stack. Thus, the execution flow continues from the address $0003 and the program execution is completed.

4.6 Time features of code execution and delay programs

The required time for executing one or more instructions is based on the main clock frequency and on the corresponding microcontroller architecture. The clock frequency is constant and from the corresponding period, the machine cycle is derived. If for example, the frequency of the main clock is 16MHz, then the clock cycle or machine cycle (MC) will be:

$$MC = \frac{1}{16MHz} = 62.5 \times 10^{-9} \text{ Sec} = 62.5 \text{ nSec}$$

The instruction execution requires one or two MCs. But how is it possible for an instruction to be executed only in one MC?

The AVR microcontrollers achieve high performance by exploiting the following features:

(a) constant instruction length which allow the whole instruction (with arguments) fetch-transfer to the execution unit (MCU) with one single operation

(b) Harvard architecture which allows the operations acceleration by using different memory for data and instructions respectively

(c) pipelining method which allows the current instruction execution and the next instruction transfer to the MCU to be performed at the same time

Figure 4.27 shows in general the operations sequence that take place for executing two instructions in a conventional system (**without Harvard architecture**). The whole operation is analyzed as follows:

(1) Fetch-Transfer of the first instruction in the execution unit (MCU)
(2) Instruction execution (result transfer to destination)
(3) Fetch-Transfer of second instruction in the execution unit (MCU)
(4) Instruction execution (result transfer to destination)

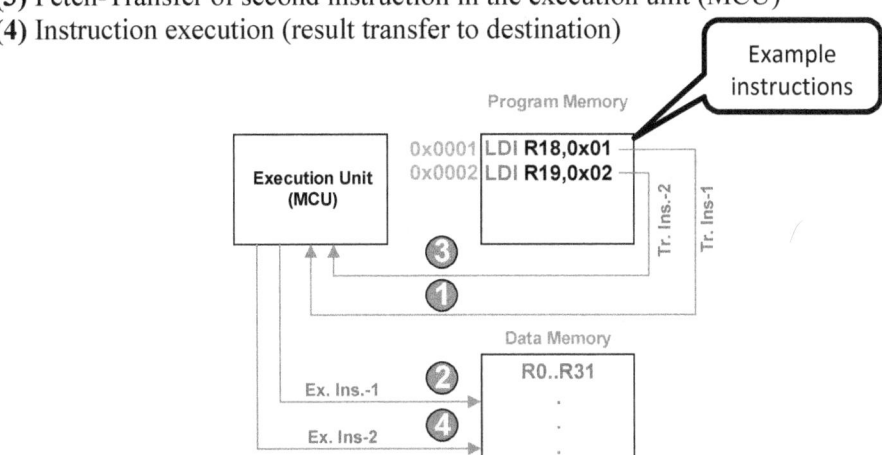

Figure 4.27 General operation of instruction execution (hypothetical)

Figure 4.28 shows the time sequence of the above operations. Every step (fetch-transfer, execution) requires one MC to be completed. Thus, the above operations (four steps), are represented by four time slices T1, T2, T3 and T4. In the same figure, the sub-operations which take place during the execution phase of each instruction are also presented.

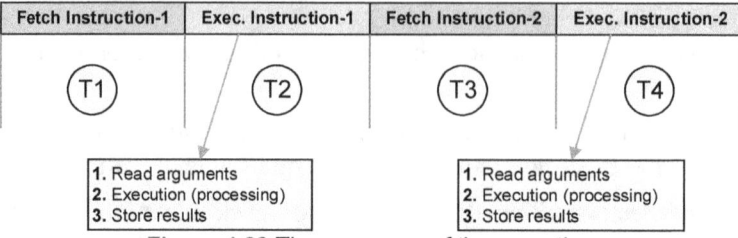

Figure 4.28 Time progress of the operation

From the above figures it is obvious that in order to start an instruction execution, the corresponding operation of the previous one has to be completed. For accelerating the instruction execution operation in the AVR microcontrollers, the pipeline technique is used. Based on that technique, there is an operation overlapping which means that two operations for two different instructions can be taken place at the same time. Figure 4.29 shows the operations that take place regarding the instruction execution:

(1) The first instruction is transferred to the execution unit (MCU)
(2) The first instruction is executed, while the second is transferred in the execution unit (MCU)
(3) The second instruction is executed, while the third is transferred in the execution unit (MCU)

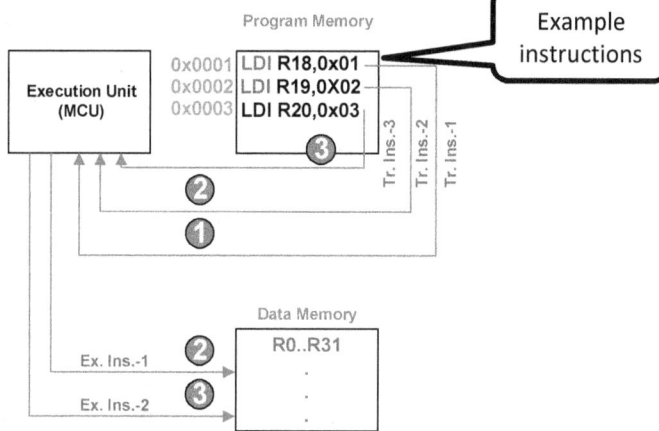

Figure 4.29 General operation of instruction execution (pipeline technique)

The time diagram of figure 4.30 shows the above process.

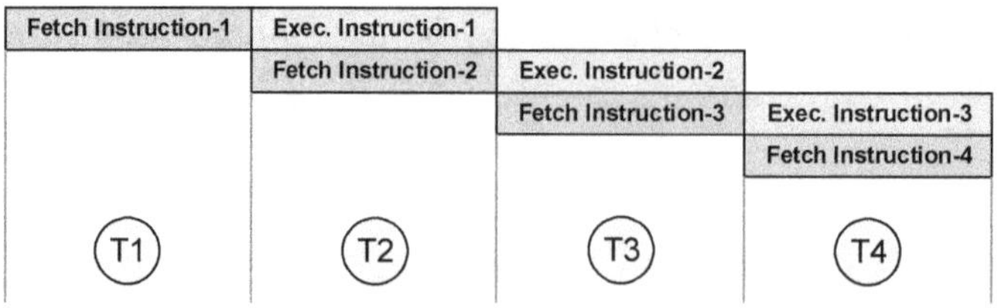

Figure 4.30 Time progress of the operation (pipeline technique)

Based on the above techniques and the architectural features, the most of the instructions of the AVR microcontroller are executed only in one machine cycle (MC).

Calculating the execution time (time delay)

In many cases there is a need for a time delay. Due to the fact that there is not any ready program for implementing time delays, it is necessary to be developed by the programmer. On the other hand, the time of the delay must be specific according to the corresponding requirements. For calculating this time delay, the required MC for every instruction execution must be known. The following code represents a time delay subroutine which is called with the name `WAIT4` (e.g. `RCALL WAIT4`). In this subroutine a loop of 100 iterations has been implemented which consists of instructions that do nothing (NOP, one MC for execution).

Code 4.11

```
WAIT4:                      ;Time delay subroutine
        LDI R21,100         ;Counter initialization
                            ;(Execution time = 1 MC)
    START:                  ;Reference point
        NOP                 ;No operation
                            ;(Execution time = 1 MC)
        NOP                 ;No operation
                            ;(Execution time = 1 MC)
        NOP                 ;No operation
                            ;(Execution time = 1 MC)
        DEC R21             ;Counter decrement
                            ;(Execution time = 1 MC)
        BRNE START          ;Condition check and return to START
                            ;Execution time = 2 MCs
                            ;for returning to START
                            ;or 1 MC for exiting from the loop
        RET                 ;Exit from the subroutine
```

Table 4.3 shows the required MCs as well as the corresponding executions iterations.

Table 4.3 Machine Cycles in the iteration

Code	Machine Cycles	Iterations
WAIT4:		
LDI R21,100	1	0
START:		
NOP	1	100
NOP	1	100
NOP	1	100
DEC R21	1	100
BRNE START	2 (True) / 1 (False)	99
RET	4	0

The instruction `LDI` is outside the loop and thus will be executed only once. The loop instructions `NOP` and `DEC` will be executed 100 times, while the `BRNE` will be executed 99 times, due to the fact that the first time the content of the register R21 is 99 and not 100. Thus, the total execution time of the code in the subroutine WAIT4, will be:

For a MC of 62.5nSec (it is assumed a clock frequency of 16MHz)
    [1+(1+1+1+1+2)x100)-1+4]x62.5nSec=
    [1+6x100-1+4]x6.25nSec=
    604x62.5nSec=
    37750 nSec=
    37.75 μSec

The instruction BRNE returns the execution flow to the label START 99 times, while the last time requires one MC due to the fact that the condition is FALSE and an exit from the loop is performed. The multiplication is performed with the 100 and not with 99. Thus, an 1 is subtracted, because the 100th time the execution flow exits from the loop and the instruction BRNE requires only 1 MC. This time delay can be expressed in mSecs and Secs, as follows:

0.03775 mSecs
0.00003775 Secs

From the above it is obvious that for achieving a time delay of 1 Sec, an accurate calculation must be performed by using multiple loops.
    For achieving accurately the desired time delay, the timers of the AVR microcontroller can be used.

4.7 Macroinstructions

The subroutines as they presented previously, they just called in order to execute the included code. On the other hand, special subprograms can be developed for representing a "new" instruction, known as macroinstruction. At the same time, there is the flexibility to pass parameters (like arguments) to the macroinstruction and thus, the corresponding operation can be adapted to current needs. The code of a macroinstruction begins with the directive .MACRO and ends with the directive .ENDMACRO. Figure 4.31 shows that the code of a macroinstruction is activated within the main code by using the corresponding name. In the figure 4.31, a macroinstruction with the name Aname has been created and can be used as a conventional instruction.

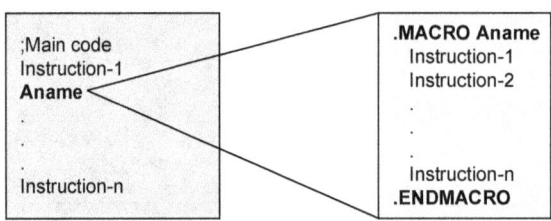

Figure 4.31 Creating and using a macroinstruction

As mentioned before, parameters can be passed to the macroinstruction for adapting the corresponding operation to current needs. The parameters are modelled with the symbolic names @0, @1, to @9 (within the macroinstruction code). In the macroinstruction usage, the proper parameters have to be placed in order the @0 to @9 to be replaced by the real values. The following code is indicative (code 4.12).

Code 4.12

```
;Place here the proper INC file for your microcontroller model
;(if needed), e.g. ATmega32/ATmega32A => m32def.inc/m32Adef.inc,
;ATmega328 => m328def.inc

.INCLUDE "include/m32def.inc"

;Macroinstruction without parameters
.MACRO testnoparam
LDI R16,0xFF
LDI R17,0x00
LDI R18,0x0D
.ENDMACRO

;Macroinstruction with three parameters
.MACRO testparam
LDI R16,@0
LDI R17,@1
LDI R18,@2
.ENDMACRO

;Macroinstruction with one parameter
.MACRO testoneparam
LDI R16,@0
.ENDMACRO

;Main code
;Using the macroinstruction without parameters
testnoparam

;Using the macroinstruction with three parameters
testparam 0xFF,0x00,0x0D

;Using the macroinstruction with one parameters
testoneparam 0xFF

;End of code
end: RJMP end
```

As shown in the code 4.12, three macroinstructions have been developed with names `testnoparam` (no parameters), `testparam` (three parameters) and `testoneparam` (one parameter) respectively. In the main code, initially, the first macroinstruction is used (`testnoparam`) without any parameter. The next macroinstruction (`testparam`) is used as follows:

`testparam 0xFF,0x00,0x0D`

134 CHAPTER 4

The three parameters (0xFF,0x00,0x0D), are stored in @0, @1 and @2:
@0=0xFF
@1=0x00
@2=0x0D

Thus, the final code that will be executed inside the macroinstruction will be:

```
LDI R16,0xFF ; @0=0xFF
LDI R17,0x00 ; @1=0x00
LDI R18,0x0D ; @2=0x0D
```

Figure 4.32 shows the macroinstruction activation from the main code. In the same figure, the parameters that are passed to the macroinstruction are also shown (the macroinstructions that they have parameters).

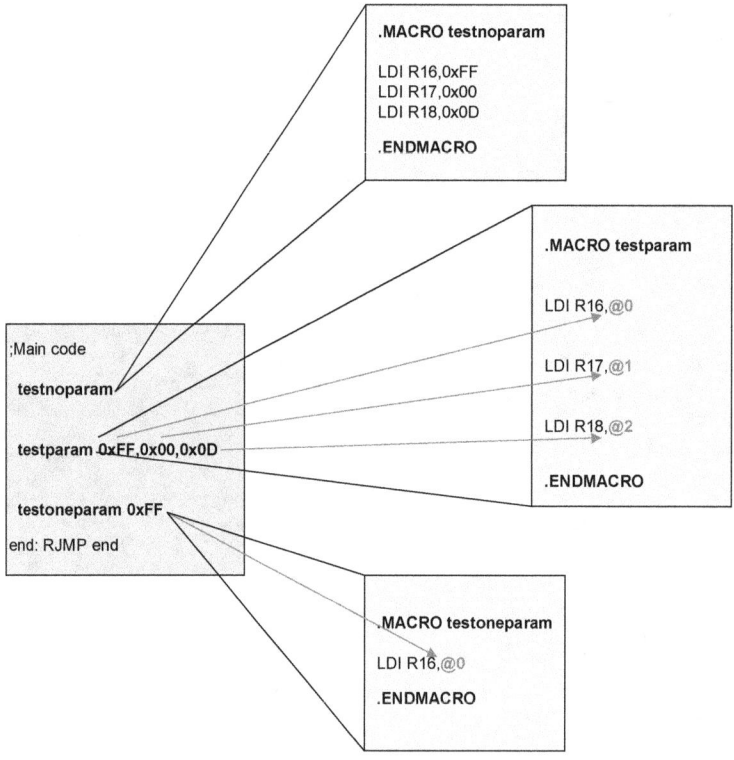

Figure 4.32 Macroinstruction "activation"

Note
It is important to notice that during the source code compilation, the macroinstruction code is reproduced every time is used.

4.8 Program coding in memory

Until now, a large part of the AVR instructions has been presented and analyzed. Moreover a more completed picture regarding the program memory organization has been created to the reader. It is also important to be recalled that every location of the program memory stores 16bits of code. For better understanding how locations and memory addresses are used, the program coding within the program memory will be studied.

As a program coding example the following code will be used:
```
LDI R16,0x01
LDI R31,0x01
```
The 16bits of the instruction `LDI Rd, k` (based on the AVR datasheet) are shown in the figure 4.33:

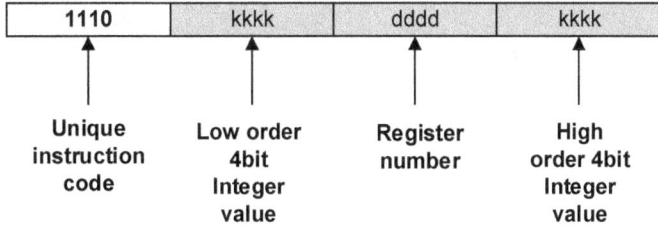

Figure 4.33 Instruction code

Figure 4.34 shows how the 16bits code for the above instructions is formed. The number 1110_2 remains constant and represents the unique code of the instruction `LDI`. The 8bits (kkkk kkkk) for the integer value that will be loaded will have the value 10, (the lower order 4bits first, 01). The next 4bits (dddd) represent the register number which is used as the first argument in the instruction (the instruction `LDI` is valid only with the registers R16 to R31). These 16 registers correspond to the numbers 0x0 to 0xF. Thus, the 4bits (dddd) for the register R16 (instruction `LDI R16,0x01`) will have the value 0 (0000_2). Finally, the code for the second instruction (`LDI R31,0x01`), differs only to the register that is used. The register R31 corresponds to the number 0xF and thus, the field dddd will has this value.

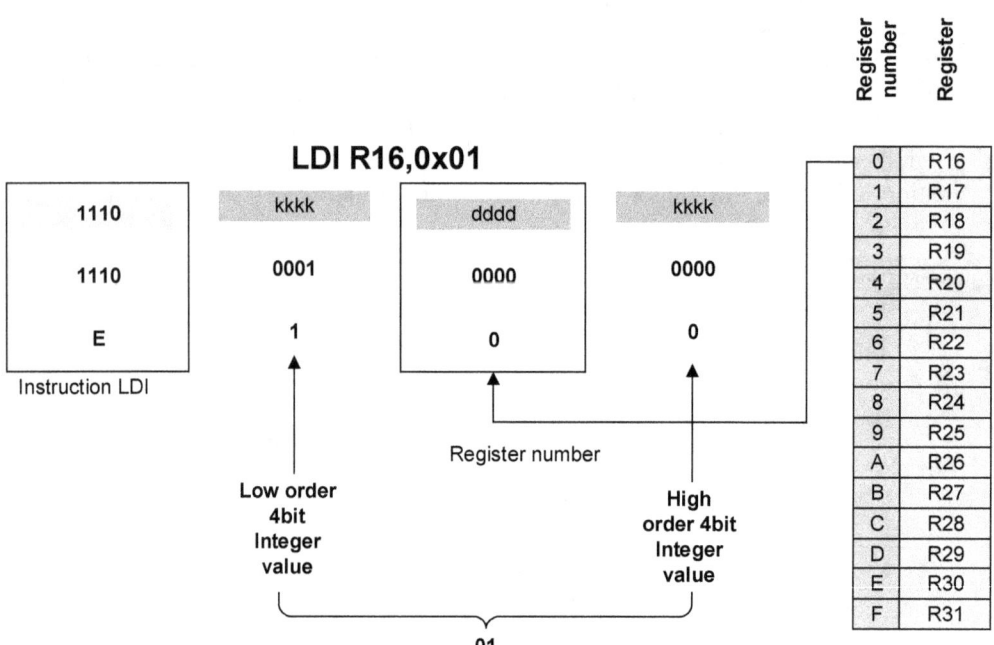

Figure 4.34a Instruction coding in memory

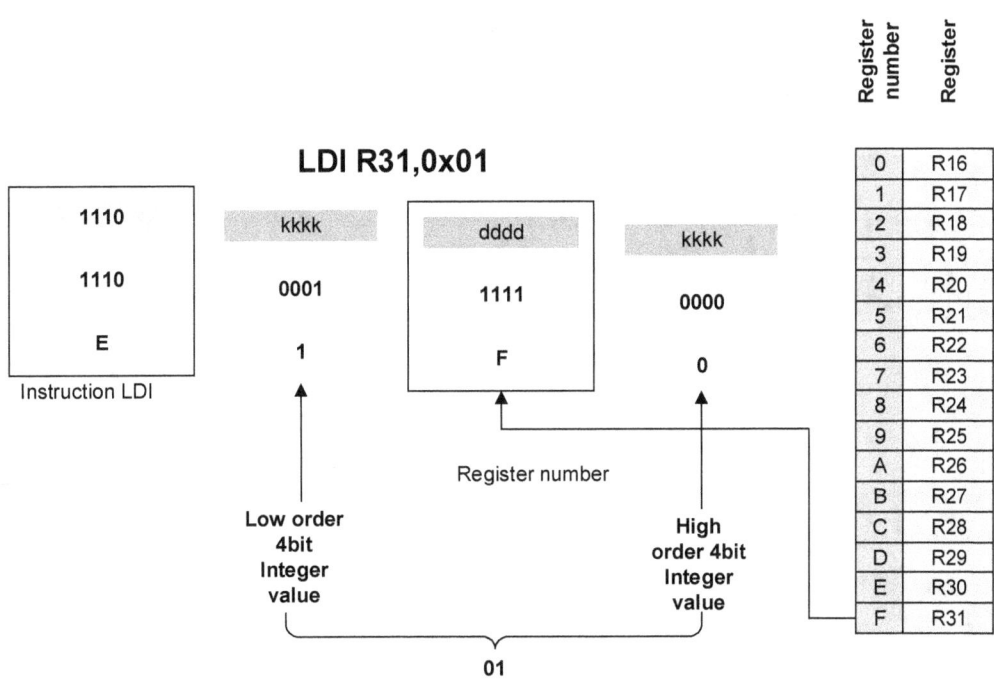

Figure 4.34b Instruction coding in memory

EXERCISE SHEETS
Implementing basic programming structures

1. Develop the proper code and the corresponding flow chart diagram for the following condition checks:

Pseudocode
If R18=0, then
        R19=1
   Otherwise
        R19=2
End-If	

Code	**Flow chart diagram**

138 ☐ CHAPTER 4

Pseudocode	

If (R18=4 and R19=5), then
        R19=0
    Otherwise
        R19=1
End-If	

Code	**Flow chart diagram**

Pseudocode	

If R18 ∈ [10,20], then
        R19=15
    Otherwise
        R19=2
End-If	

Code	**Flow chart diagram**

2. Develop a program in order to store in the memory addresses $100 to $200 the number 0xFF.

Address	Content
$100	FF
$101	FF
....	....
....	....
....	....
$1FF	FF
$200	FF

Code

3. Develop a program to exchange the first half and the second half of the memory area $100 to $200.

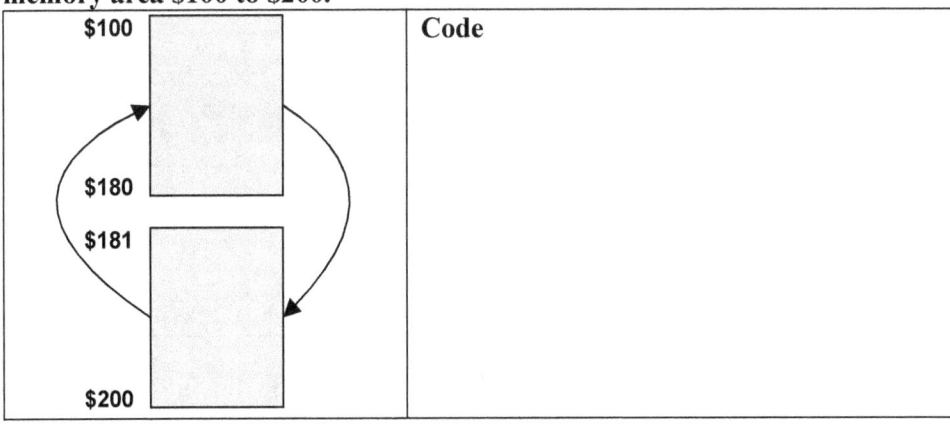

Code

4. Develop a program in order to calculate the number of zero and positive values of an array which is stored in the memory locations $40 to $80. The number of the zero values will be stored in the register R20, while the number of the positive values will be stored in the register R21.

5. Develop a program for finding the minimum value within an array of 16 locations (starting from the address $80). For the corresponding implementation, use the following pseudocode:

```
MIN=matrix [$80]
For A=$81 to $8F
        If matrix[A]<MIN, then
                MIN=matrix[A]
        End-If
```

6. Develop a program for reading the contents of an array which is located in the memory area $100-$1FF and replacing all the zero values with the number 0xFF.

7. Develop the proper programs for implementing the following algorithms:

If (R18≥10 and R18≤20), then
    R20=100
Otherwise
    R20=200
	If R18<10, then
	    R20=200
	Otherwise
	    If R18>20, then
	        R20=200
	    Otherwise
	        R20=100
	    End-If
	End-If

Code	Code

8. Develop a program for copying 20 bytes starting from the address $040 to another memory area which starts from the address $1D0.

9. Develop a program in order to check if (R18+R19+R20)>50. If this is TRUE, the bits 7, 5 and 1 of the register R21 will be cleared (without affecting the rest of the bits).

10. It is initially assumed that the content of the register R16 represents the status of eight sensors in a control system.

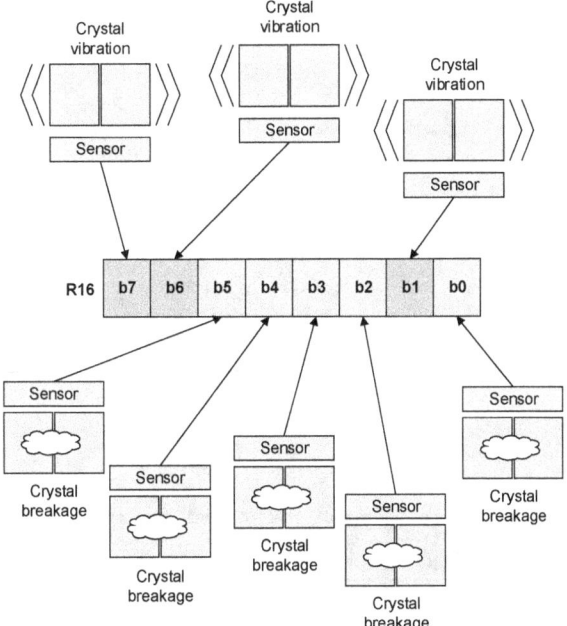

The sensors detect crystal vibration or breakage in a sensitive product area. The vibration detection is critical (warning alarm) in order to prevent the crystal breakage (red alarm). Based on a warning alarm, the personnel tries to solve the issues. The registers R17 and R18 represent the activation of the warning and the red alarm respectively (value of activation 0xFF). The bits 7,6 and 1 of the register R16 are activated from the sensors regarding the warning alarm, while the rest of the bits are activated in the case of a red alarm. Develop a program for detecting the sensors status in order to activate the warning or the red alarm.

11. In a building there is installed an automated system for watering lawns and ornamental plants. The watering has to be performed at specific days and hours. The watering will be different for different plants and thus, different settings have to be used.

The watering will be performed as follows:

Ornamental plants
The months 1,3,5,7,9 and 11
The days 10 and 20 of the above months
The watering day, the system will be activated at 4.00pm and will be closed at 5.00pm

Lawns
The system will be activated every day at 8.00pm and will be closed at 9.00pm

It is also assumed that the microcontroller is connected with a real time system in order to read the real date and time. The date and time data are stored in general purpose registers as follows:

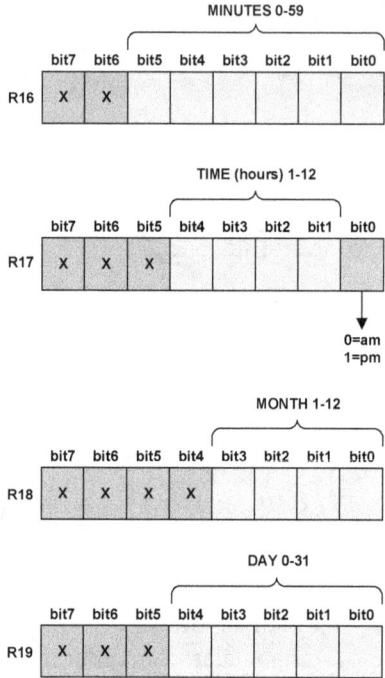

Develop a program in order to check the above registers contents and to activate the two watering systems as follows:
Ornamental plants watering (R20=0xFF=activation), (R20=0x00=closing)
Lawns watering (R21=0xFF=activation), (R21=0x00=closing)

12. Develop a program to calculate:

(a) $S=1+3+5+\ldots+10$
(b) $S=1^2+2^2+3^2+\ldots+10^2$
(c) $S=(n-1)^2+(n-2)^2+\ldots+(n-10)^2$ for $n=10$

13. It is assumed that the register R20 represents the status of the background of eight lighted traffic signs. Develop a program to invert continuously the background 20000 times.

Implementing basic programming structures 143

Indicative operation of the
electronic traffic sign

Background=1   Background=0

14. In a tunnel of the national highway there are 3 locations of alarm activators for the drivers and 3 locations with gas detectors. Moreover, there are 2 locations of escape paths. Based on the tunnel management rules, in a case of emergency (activated alarms), the closer escape path must be lighted with a proper sound alarm for helping drivers and passengers to be driven at the emergency exit.

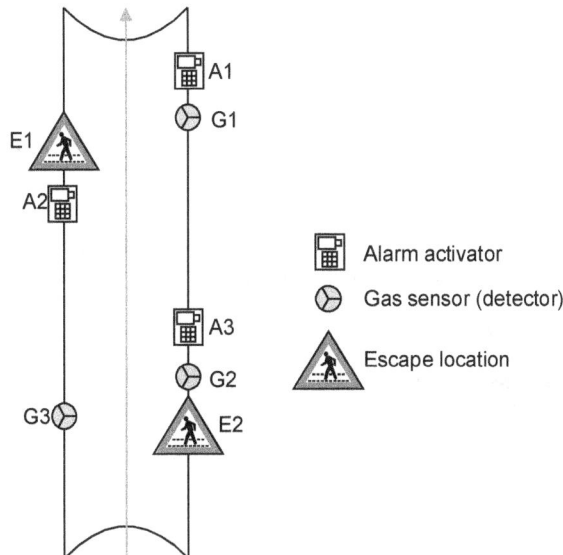

The logic of the tunnel management system can be described as follows:

If G1=1 or A1=1 or A2=1, then E1=1 (0=no active, 1=active)
If G2=1 or G3=1 or A3=1, then E2=1 (0=no active, 1=active)
Of course, the activation of E1 and E2 may be happened at the same time. The gas detection is performed by sensors, while the alarm can be activated by drivers using a special red button. The buttons are install in a communication device which can be used also by drivers in a case of emergency.
The register R16 represents at any time the status of the G1, G2, G3, A1, A2 and A3. The above status is mapped on specific bits as follows:

	7	6	5	4	3	2	1	0
R16	X	X	G3	G2	G1	A3	A2	A1

Develop a program for detecting the tunnel status (R16 content) and activating the escape paths as follows:

Bit7-4 (R17): Sound signal (alarm) E1
Bit3-0 (R17): Light sign E1

Bit7-4 (R18): Sound signal (alarm) E2
Bit3-0 (R18): Light sign E2

15. Develop a triple loop for 100000 iterations in total.

16. Use the stack in order to implement a triple loop for 1000 iterations in total using the same register as counter.

17. If the main clock frequency is 1MHz, calculate the total execution time for the programs of the exercises (15) and (16).

18. Calculate the total execution time of the code 4.11, if the main clock frequency is 1MHz.

19. In a previous exercise, a triple loop has been implemented. Develop a macroinstruction in order to pass from the main code, three parameters for the initial values of the corresponding counters.

20. Using the AVR datasheet, describe and analyze the memory coding of the instruction `ADD R17,R18`.

5 Basic Programming of the Input/Output (I/O) Ports

Content-Goals
The I/O ports constitute the most important part of the microcontroller, due to the fact that allow operation control and support communication with external devices. In this chapter, the main features of the I/O ports as well as the corresponding operation will be presented.

Chapter Contents
5.1 Introduction
5.2 Elementary electrical-electronic circuits
5.3 Basic manipulation of digital ports
5.4 LED manipulation through microcontroller ports
5.5 General purpose switch circuits
5.6 Implementing a time delay

5.1 Introduction

The access to microcontroller hardware and the corresponding programming is based on registers. The microcontroller communication with the external environment (e.g. external circuits, sensors) is achieved through the ports which recognize analog or digital signals (analog signals are recognized using dedicated pins in specific microcontroller models). Moreover, the I/O port control is based on special registers known as I/O registers. When the contents of an I/O register is changed, the corresponding pin status is also changed. Thus, a "direct connection" exists between I/O registers and I/O pins.

Note
The digital I/O ports can be used for simple operations (e.g. LED activation, read a switch) or more complex using signals and protocols.

5.2 Elementary electrical-electronic circuits

5.2.1 Circuit with resistor

Before the explanation of controlling external circuits through the I/O ports, it is very important the analysis of some elementary electrical circuits for understanding the basic concepts.
It is known that if a voltage source and a resistor is used, then an electrical circuit can be formed with the corresponding current flow. Figure 5.1 shows a basic electrical circuit which consists of a voltage source and a resistor.

146 CHAPTER 5

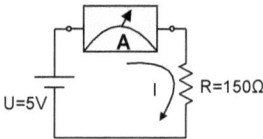

Figure 5.1 Basic electrical circuit

In the circuit of figure 5.1, a current I flows which is limited from the resistor R. The current I is calculated as follows:

$$I = \frac{U}{R} = \frac{5V}{150\Omega} = 0.0333A = 33.3mA$$

The current I can be also measured by using an ampere meter (fig. 5.2)

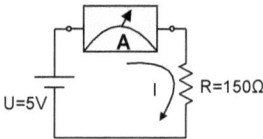

Figure 5.2 Current measurement in the circuit

If the voltage U_R of the resistor is measured using a volt meter (fig. 5.3), then the U_R will be:

$$U_R = 5V$$

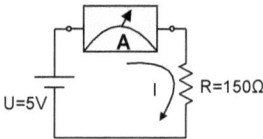

Figure 5.3 Voltage measurement at the resistor

Additionally, if two identical resistors of 75Ω are used in the circuit (fig. 5.4), then the current I will be calculated as follows:

$$I = \frac{U}{R_1 + R_2} = \frac{5V}{150\Omega} = 0.0333A = 33.3mA$$

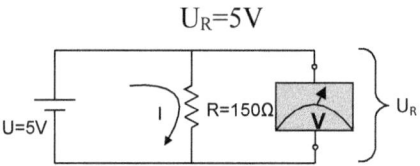

Figure 5.4 Circuit with two resistors

The sum of the resistor voltages is equal to the source voltage (fig. 5.4):

$$U = U_{R1} + U_{R2}$$

If the current I is known, then the voltage drop in each resistor will be:

$$U_{R1} = I * R_1 = 33.3mA * 75\Omega \cong 2.5V$$
$$U_{R2} = I * R_2 = 33.3mA * 75\Omega \cong 2.5V$$

The circuit of figure 5.4 constitutes a voltage divider, where the source voltage is divided by two (when the resistors are identical). Such a circuit is very useful when the adaptation of an external signal (which exceeds 5V) has to be applied for implementing the connection with a microcontroller pin. Figure 5.5 shows that the two circuits (with 150Ω and 2x75Ω resistors) are equivalent.

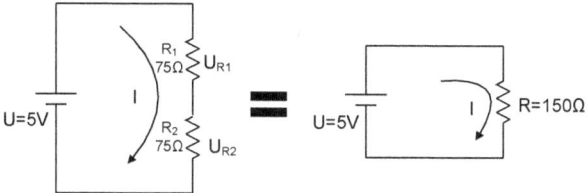

Figure 5.5 Equivalent circuits

Thus, the same current I flows in both circuits of figure 5.5 (R=R1+R2):

$$I = \frac{U}{R} = \frac{U}{R_1 + R_2} = \frac{5V}{150\Omega} = 0.0333A = 33.3mA$$

5.2.2 Circuit with diode

The diode is an electronic element (semiconductor) which allows the current flow when is biased correctly (forward). Figure 5.6 shows the diode symbol.

Figure 5.6 Diode symbol

As shown in figure 5.6, the diode has a positive (anode) and a negative (cathode) pin. Based on the pin connections, the current flow is allowed or not. The diode behavior diagram is shown in figure 5.7. For the forward bias (positive diode pin to the positive voltage and negative diode pin to the negative voltage or ground) and diode voltage (pin to pin) greater than V_F, a positive current flow is allowed. The V_F voltage is depended on the construction material (usually 0.7V for Si-Silicon and 0.3V Ge-Germanium). Additionally, for high reverse voltages the diode collapse is happened allowing high reverse current flow. In the case of microcontrollers, the above situation is not faced by the engineer.

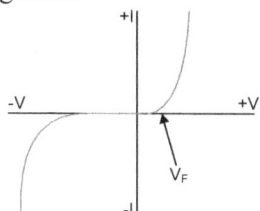

Figure 5.7 Diode behavior diagram

Figure 5.8 shows the current flow using a basic electrical circuit with a resistor and a diode.

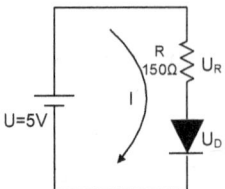

Figure 5.8 Circuit with a resistor and a diode

Due to the voltage source, a current flows in the circuit of the figure 5.8. On the other hand, the diode resistor is not known and thus, the current will be calculated using the voltage drop across the resistor.
For the circuit voltages it is known that

$$U = U_R + U_D$$

Thus, the voltage drop across resistor is

$$U_R = U - U_D$$

The circuit current can be expresses as

$$I = \frac{U_R}{R} = \frac{U - U_D}{R}$$

If it is assumed that the circuit has a silicon (Si) diode, then the voltage drop across diode is $U_D = 0.7V$. Thus, the final current of the circuit is:

$$I = \frac{5V - 0.7V}{150\Omega} = \frac{4.3V}{150\Omega} = 0.0286A = 28.6mA$$

5.2.3 The LED diode

The LED (Light Emitted Diode) is a special diode which emits light when is biased forward. Figure 5.9a shows the symbol of the LED.

Figure 5.9a LED symbol

The physical form of five LEDs with different colors is shown in figure 5.9b.

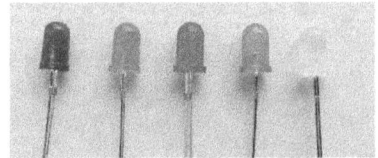

Figure 5.9b 5mm LEDs with different colors

The LED diode has a positive and a negative pin, and works like the classic diode.

Note

A main difference of the LED diode as compared to classic one, is that the corresponding voltage drop across positive and negative pins is much greater. For a red LED, this voltage is 1.8 to 2.0V. Thus, it is assumed that $U_D=2V$, when a current calculation takes place.

Figure 5.10a shows a LED diode in a simple electrical circuit. For the same circuit, the voltage source can be replaced by a microcontroller pin (fig. 5.10b). Thus, the LED can be controlled by the microcontroller based on the corresponding software. It must be noticed that, due to the small LED resistance, an additional resistor must be used for limiting the current. Otherwise, the LED and the microcontroller port can be damaged.

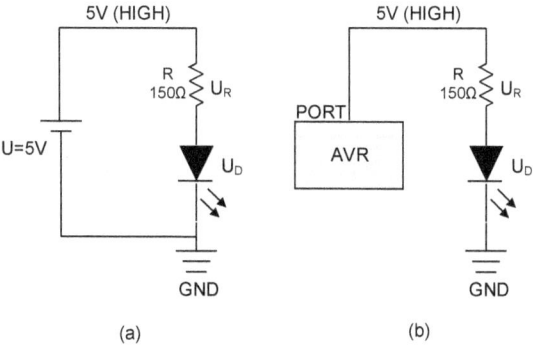

Figure 5.10 LED control from a voltage source or from a microcontroller

The current of circuit 5.10 is:

$$I = \frac{U - U_D}{R} = \frac{5V - 2V}{150\Omega} = \frac{3V}{150\Omega} = 0.02A = 20mA$$

Figure 5.11 shows how the LED can be lit by setting the microcontroller pin level (e.g. PB0) to HIGH (5V) or LOW (0V).

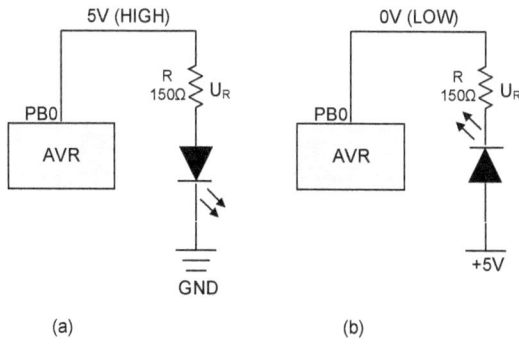

Figure 5.11 LED activation with a signal 5 or 0V

In figure 5.11a, the pin PB0 is HIGH for lighting the LED. On the other hand, PB0 (fig. 5.11b) is LOW for lighting the LED which is connected in the opposite direction (now, the negative voltage is applied on PB0).

5.3 Basic manipulation of digital ports

As shown in figure 5.12, there are three I/O port registers. The subscript x corresponds to the symbolic name of the port. For example, the registers of port B are defined as DDRB, PORTB and PINB. Based on the desired operation, the proper register is used. Thus, every operation (e.g. data direction and data read/write) is distinct and is implemented through a different register.

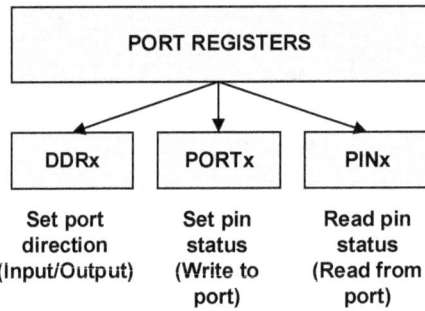

Figure 5.12 Port registers

Table 5.2 shows the special I/O registers for each port of the microcontroller.

Table 5.1 Special I/O port registers

Port	Register	Operation
A	DDRA	Data direction (Input/Output)
	PORTA	Write data (output)
	PINA	Read data (input)
B	DDRB	Data direction (Input/Output)
	PORTB	Write data (output)
	PINB	Read data (input)
C	DDRC	Data direction (Input/Output)
	PORTC	Write data (output)
	PINC	Read data (input)
D	DDRD	Data direction (Input/Output)
	PORTD	Write data (output)
	PIND	Read data (input)

Note
The number of ports is different based on the microcontroller model. Additionally, in the .inc file (e.g. m328def.inc for ATmega328 or m32def.inc for ATmega32) the mapping between physical addresses and symbolic names of the ports is included. For example, the register PORTB can be used for different microcontrollers (e.g.

ATmega328 and ATmega32), despite the fact that the corresponding physical addresses are different (e.g. $05 and $18).

Before any port operation, the data direction must be set (from or to the microcontroller). This is achieved through the DDRx register, where x is the symbolic name of the port.

As shown in figure 5.13, each pin can be defined as input or output separately. The value 1 corresponds to the output direction, and the value 0 to the input direction. In the example of figure 5.13, the register DDRx (x=symbolic name of the port) is used for defining the data direction of each pin through the 8bit value.

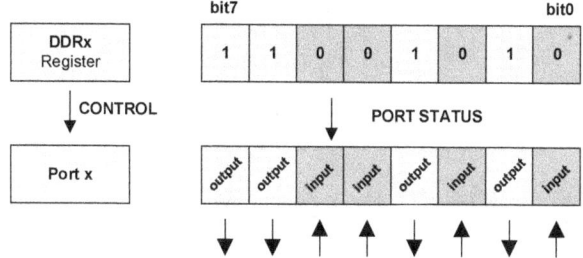

Figure 5.13 Register DDRx usage

In the above example (figure 5.13) the bit value 11001010 is loaded in the register DDRx.

OUT – Write to an I/O register

A special instruction is needed for loading a value in an I/O register. More precisely, the loading operation in an I/O register is implemented by the instruction

```
OUT IOReg, Rs
```

where `IOReg` is the I/O register (destination) and `Rs` a general purpose register (R0 to R31). The `IOReg` can be expressed with the corresponding symbolic name or the port address (0 to 63 or $0 to $3F).

The sequence 11001010 (0xCA) can be loaded in register DDRA by writing

```
LDI R16,0xCA
OUT DDRA,R16
```

or

```
LDI R16,0xCA
OUT $1A,R16   ;For a specific microcontroller model
```
using the port address.

Figure 5.14 shows the "connection relation" of the DDRB register with the port B pins. This figure corresponds to the ATmega328 model and shows also the direct loading of the binary number to the temporary register R16. The pin order based on the available ports is different according to the microcontroller model.

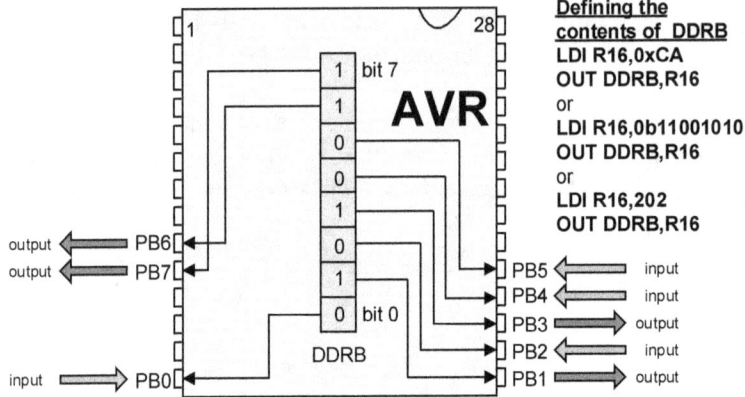

Figure 5.14 Port B pins in the ATmega328

Using a different microcontroller model (e.g. ATmega8515 or ATmega32), the pins of port B have a different order in the chip (fig. 5.15).

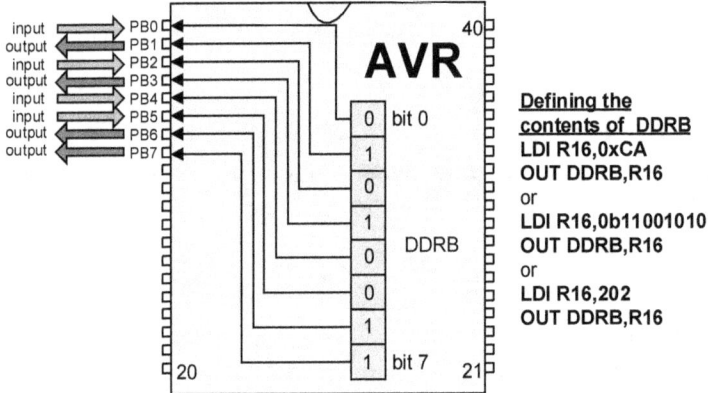

Figure 5.15 Port B pins in ATmega8515 and ATmega32

Note
The port manipulation methodology is the same for all the AVR microcontrollers. In many examples, indicative microcontrollers have been used due to the fact that have been implemented in real circuits (for the purposes of the book) and also because a practical point of view had to be presented for giving a clue to the reader for the real circuits.

After the data direction set (through the DDR register), the real data can be sent by using the proper register. The data transmission through the output pins of the microcontroller is achieved by using the register PORTx (x=name of the port). The data transmission (e.g. binary number 11011000) to the outer environment of the microcontroller is achieved by register PORTx, assuming that all the pins of port B have been set as outputs with the instructions `LDI R16,0xFF` and `OUT DDRB,R16`.

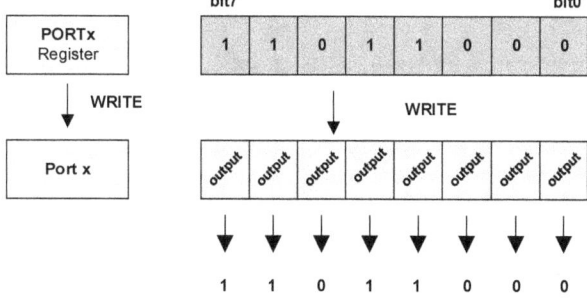

Figure 5.16 Using the register PORTx

As shown in figure 5.16, anything loaded in PORTx, is appeared as voltage level to the corresponding pins. The complete code for writing the binary number 11011000 in port B, is as follows:

Code 5.1

```
LDI R16,0xFF          ;load 11111111 in R16
OUT DDRB,R16          ;set all the pins of port B, as outputs
LDI R16,0b11011000    ;load 11011000 in R16
OUT PORTB,R16         ;write the data in the pins of port B
```

Figure 5.17 shows the data direction set for the port B through register DDRB as well as the pin voltage set (binary number 11011000) through register PORTB. **This operation is the same for all the AVR microcontrollers.** In the same figure, the pinout of the ATmega8515 and ATmega32 is used.

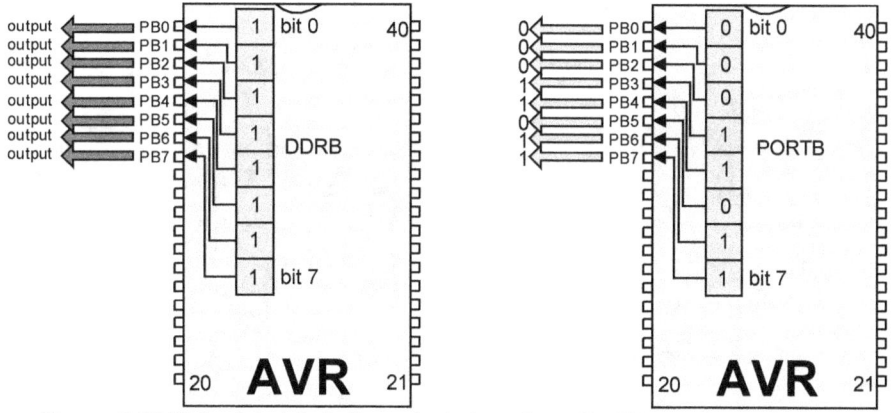

Figure 5.17 Write data bits in the output pins of port B (ATmega8515/ATmega32)

IN – Read from an I/O Register

For reading from an I/O register, the following instruction is used:

```
IN Rd,IOReg
```

Where `Rd` is a general purpose register (R0 to R31) and `IOReg` the I/O register (source). For using the I/O register, the symbolic register name can be used as well as the corresponding address (0 to 63 or differently $0 to $3F).

The *read* operation from register `PINB` (applied signals in port B pins) can be performed using the instruction

`IN R16,PINB`

or

`IN R16,$16   ;for a specific microcontroller model`

If it is assumed that the binary number 11011000 represents the voltage levels of the port x pins, then the *read* operation will be performed through register PINx (x=name of the port). Figure 5.18 shows that the signals (voltage levels) of the input pins can be read through the contents of PINx register.

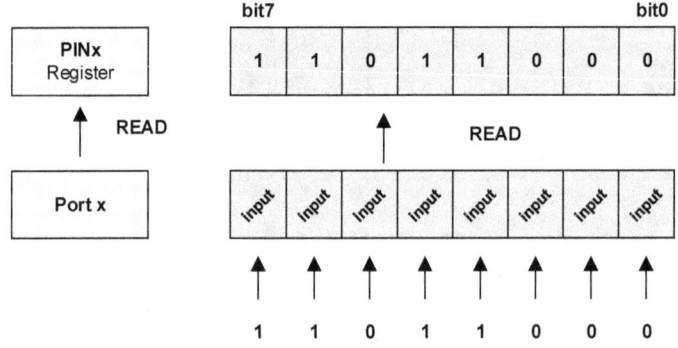

Figure 5.18 Using PINx register

The *read* operation from port B for the ATmega8515 and ATmega32 microcontrollers is shown in figure 5.19.

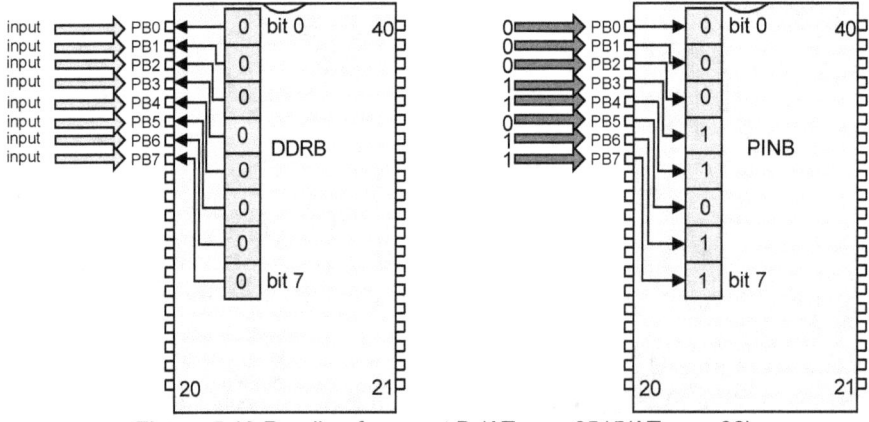

Figure 5.19 Reading from port B (ATmega8515/ATmega32)

The complete code for reading from port B (port pins), is as follows:
Code 5.2

```
LDI R16,0x00   ;Load 00000000 in R16
OUT DDRB,R16   ;Set all the pins of the port B as inputs

IN  R16,PINB   ;Read from port B and store input data in R16
```

Note
The logical signals 0 and 1 represent the voltage levels 0V and 5V respectively. When the signals are applied to an external circuit, a capable current is flown for lighting a LED or controlling a device (30-40mA maximum current per pin-typical value).

bit manipulation in an I/O register
The manipulation of specific bits in I/O registers, allows the control of specific pins simplifying the corresponding operation.

SBI - Set bit (value=1) in an I/O register
In many cases, is necessary to access a specific bit of an I/O register. For example, a selected pin of port B has to be set as output or only a selected LED has to be lit without changing the rest of the pins. With the instruction

```
SBI IOReg,b
```

with IOReg ∈ [0,31] (access to the first 32 I/O registers) and b ∈ [0,7]

the bit b of the `IOReg` register is activated (set, logical 1)

Example
For lighting the connected LED to pin PB0 (pin 0 of port B), the following instruction has to be given:

```
SBI PORTB,0 or SBI $18,0  ;for a specific AVR model
```

Figure 5.20 shows the bit0 activation of `PORTB` register which is used for the writing operation (output) to port B.

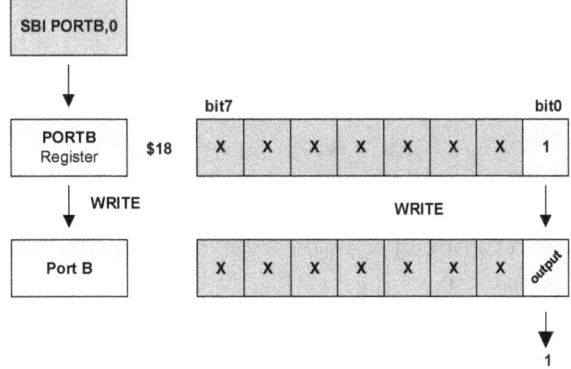

Figure 5.20 Write to pin PB0 (X = pin which is not affected)

CBI - Clear bit (value=0) in an I/O register
A specific bit of an I/O register can be cleared using the instruction

```
CBI IOReg,b
```
with $IOReg \in [0,31]$ (access to the first 32 I/O registers) and $b \in [0,7]$

Example
In order to turn off a LED which is connected to pin PB0 (pin 0 of port B), the following instruction will be used:
```
CBI PORTB,0 or SBI $18,0 ;for a specific AVR model
```

The clearance of bit0 of register PORTB which is used for writing (output) in port B, is shown in figure 5.21.

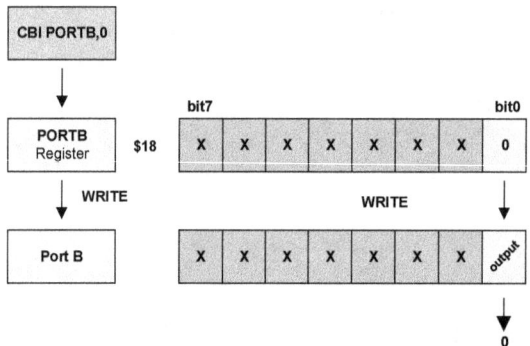

Figure 5.21 Write to pin PB0 (X = pin which is not affected)

SBIS – Scanning bit activation (value 1) in an I/O register and instruction bypass
For checking the status of specific bits and controlling the corresponding code execution flow, the following instruction is used:
```
SBIS IOReg,b
```
with $IOReg \in [0,31]$ (access to the first 32 I/O registers) and $b \in [0,7]$

This instruction bypasses the execution of the next instruction, if the bit b of register IOReg is activated.

SBIC – Scanning bit deactivation (value 0) in an I/O register and instruction bypass

For bypassing the next instruction in the case of a specific bit deactivation, the following instruction is used:
```
SBIC IOReg,b
```
with $IOReg \in [0,31]$ (access to the first 32 I/O registers) and $b \in [0,7]$

5.4 LED manipulation through microcontroller ports

As mentioned before, a LED can be connected with two different ways (different direction-bias). This means that the LED will be lit based on a HIGH or LOW signal. Figure 5.22 shows two LEDs which are connected in different directions (biases). The positive LED pin is connected to pin PB0, and the negative pin to ground (GND) through a resistor. Thus, this LED will be lit if a 5V (HIGH-logical 1) signal is applied to PB0 pin. On the other hand, the second LED is connected to PB1 through the negative pin, while the positive pin is connected through a resistor to 5V. This means that the second LED will be lit when a LOW signal (0V-logical 0) is applied to the pin PB1.

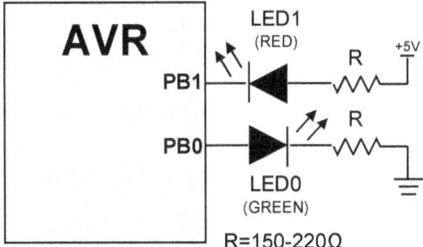

Figure 5.22 Two different LED connections

In the first example, the LEDs LED0 and LED1 will be lit (fig. 5.22). For achieving this, the first step is to set the pins PB0 and PB1 of port B as outputs. The second step is to activate the needed signals. The whole procedure will be implemented in the following steps:

(a) Pins 0 and 1 of port B (PB0, PB1) will be set as outputs. This is achieved by writing 1 in the bits 0 and 1 of the register DDRB using the instructions

```
SBI DDRB,0    ;Set bit 0 (activation, value 1)
              ;of the register DDRB (PB0, output)

SBI DDRB,1    ;Set bit 1 (activation, value 1)
              ;of the register DDRB (PB1, output)
```

(b) Activation of LED0. The positive pin of LED0 is connected to PB0 and thus, the value 1 (HIGH) will be written to bit 0 of the register PORTB. This will be achieved with the instruction

```
SBI PORTB,0   ;Set bit 0
              ;of register PORTB (PB0, HIGH)
              ;for the first LED (LED0)
```

(c) Activation of LED1. The negative pin of LED1 is connected to PB1 and thus, the value 0 (LOW) will be written to bit 1 of the register PORTB. This will be achieved with the instruction

```
CBI PORTB,1   ;Clear bit 1
              ;of register PORTB (PB1, LOW)
              ;for the second LED (LED1)
```

5.5 General purpose switch circuits

5.5.1 Operation methods

Switches and buttons (momentary switches) represent very important components of any electronic circuit, due to the fact that support user input, operation selection, procedure activation, etc. The basic issues that must be faced effectively concerning the switch operation control by the microcontroller are:

(a) the connection method with the microcontroller
(b) the signal type generating by a switch
(c) how the microcontroller detects the change of a switch status
(d) special characteristics of a switch due to the mechanical construction

Figure 5.23a shows two switch circuits known as Pull Down and Pull Up circuits. In the Pull Down circuit, when the switch is open (no current flow) the output voltage U_{out} (default output) is 0V (logical 0). On the other hand, in the Pull Up circuit, the default output (switch open) is 5V (logical 1). Table 5.2 shows the operation logic as well as the signals generating by the switch circuits of figure 5.23a.

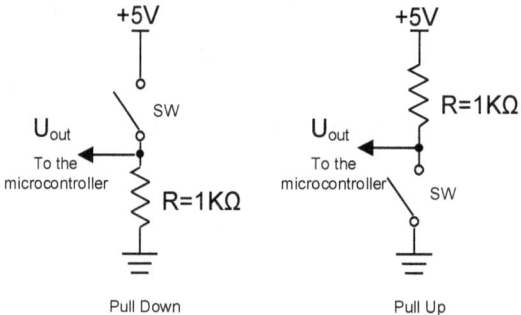

Figure 5.23a Switch circuits

Table 5.2 Switch circuit operation

	Pull Down			Pull Up	
SW	U_{out}	Logical level	SW	U_{out}	Logical level
A	0V	0	A	5V	1
K	5V	1	K	0V	0

A = switch open, K = switch closed

Figure 5.23b shows the buttons that can be used in a breadboard or in a printed circuit board (PCB). The physical form of the switch is shown in figure 5.23c. Finally, figure 5.23d shows a large button that can be used in the case of a device.

Figure 5.23b Buttons   Figure 5.23c Switch   Figure 5.23d Large button

5.5.2 Simplified switch circuit

An internal Pull-Up resistor is connected in every AVR pin. Thus, any pin can be set to HIGH (5V) without any connected external circuit. Based on the above operation, an external simple switch without resistor and voltage source can be used to drive the pin voltage level to ground (GND). Now the external circuit has only a switch component. The above technique is very useful when many switches are needed. On the other hand, must be noticed that the above switch circuit simplification can be used only when it is supported by the microcontroller. In general (for any application), the Pull Up and Pull Down switch circuits are used.

The internal Pull Up resistor of the microcontroller (e.g. for port B) is shown in figure 5.24. Figure 5.24 shows also the two different states of an external switch connected to PB0 (switch open and closed).

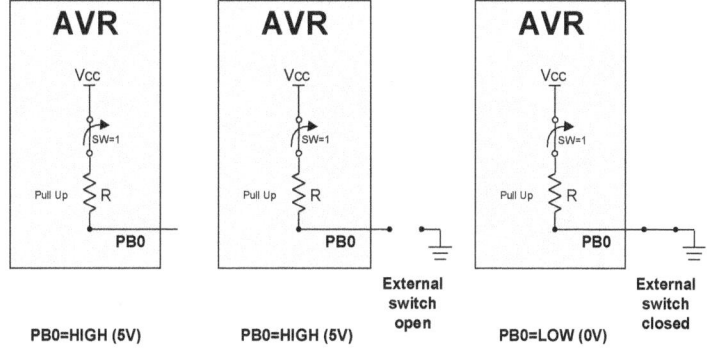

Figure 5.24 Using a switch to the ground (GND)

As shown in figure 5.24, when the internal Pull Up resistor is activated, the pin PB0 (as an example) is set to HIGH (5V). This voltage level is not changed while the external switch is open. When the external switch is closed, then the pin PB0 is driven to LOW (0V). When the external switch is open again, then the voltage level returns to HIGH (5V). Every time the switch is closed (LOW, PB0), it is detected by the corresponding code which is running within the microcontroller.

5.6 Implementing a time delay

In many cases an observable delay by the user is needed. For example, in a successive LED series activation, a time delay of 1 Sec before each activation gives to users the time to observe the whole process. In this section the code for generating a time delay of 1 Sec will be analyzed. In the following calculation, it is assumed that the microcontroller is operating at the frequency of 1MHz (default frequency in many models when an external clock crystal is not applied). The Machine Cycle (MC) for this frequency is

$$MC = \frac{1}{1MHz} = 1 \times 10^{-6} \text{ Sec} = 1 \text{ μSec}$$

Based on the previous calculation, for a delay of 1 Sec, 1 million MCs will be needed (execution of 1 million instructions of 1 MC). The registers are 8 bit (max value 255) and thus, a triple loop will be implemented. The triple loop is shown in figure 5.25. In the delay calculation of figure 5.25, the one MC at every loop exit has not be taken in account for simplifying the corresponding calculations.

```
                    LDI  R16,25    ← (1 MC)
                    Start1:
                        LDI  R17,100   ← (1 MC)
                        Start2:
                            LDI  R18,100   ← (1 MC)
1007600  40300              Start3:
cycles   cycles  400 cycles     NOP        ← (1 MC)
                                DEC  R18   ← (1 MC)
                                BRNE Start3 ← (2 MC, 1 MC at Exit)
                            DEC  R17       ← (1 MC)
                            BRNE Start2    ← (2 MC, 1 MC at Exit)
                        DEC  R16           ← (1 MC)
                        BRNE Start1        ← (2 MC, 1 MC at Exit)
```

Code 5.3

Figure 5.25 Delay code for 1 Sec at 1MHz

For developing the following example applications, some code settings and circuit simplifications will take place (see the notes 1,2 and 3).

Note 1
In every program the stack initialization takes place.

Note 2
Remember the inclusion of the proper INC file (when is needed), according to the microcontroller model which is used, in order to use symbolic names and not physical addresses inside the code.

Note 3
For simplicity reasons in the following circuits, the external circuits for Reset and Power Supply (+5V, GND) are not presented and it is assumed that already exist. Thus, each circuit is focused on the main operation of the corresponding application. Additionally, it is assumed that an external clock crystal does not exist (application is based on the main internal clock of 1MHz).

Application 5.1 – Activating eight LEDs

If the activation of multiple LEDs is needed, the general purpose registers are used (e.g. instead of setting specific bits by using the SBI instruction). Figure 5.26 shows a circuit with 8 connected LEDs.

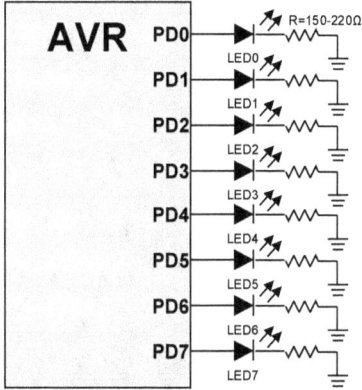

Figure 5.26 8 LEDs connected to port D

For lighting all the LEDs, the following code will be used:

Code 5.4

```
;Place here the proper INC file for your microcontroller model
;(if needed), e.g. ATmega32/ATmega32A => m32def.inc/m32Adef.inc,
;ATmega328 => m328def.inc

.INCLUDE "include/m32Adef.inc"

LDI  R16,0xFF         ;Load the value 11111111 in
                      ;register R16
OUT  DDRD,R16         ;Set pins of port D as outputs

OUT  PORTD,R16        ;Activate all pins/LEDs

END: RJMP END
```

Application 5.2 – 8 LED activation and deactivation with time delay

In this application 8 LEDs (fig. 5.26) will be turned on and off using a time delay between states.
The complete LED control code is as follows:

Code 5.5

```
;Place here the proper INC file for your microcontroller model
;(if needed), e.g. ATmega32/ATmega32A => m32def.inc/m32Adef.inc,
;ATmega328 => m328def.inc

.INCLUDE "include/m32Adef.inc"
```

```
;**************************
;Stack initialization
;**************************
LDI R16,HIGH(RAMEND)
OUT SPH,R16
LDI R16,LOW(RAMEND)
OUT SPL,R16

.DEF off=R20
.DEF on=R21


;**************************
;Main program
;**************************
LDI off,0x00        ;Load R20=00000000
LDI on,0xFF         ;Load R21=11111111
OUT DDRD,on         ;Set all the pins of port D as outputs

Again:
  OUT PORTD,on              ;Activate all LEDs
  RCALL delay_1sec          ;Delay 1 Sec
                            ;(call subroutine)
  OUT PORTD,off             ;Clear (deactivation) LEDs

  RCALL delay_1sec          ;Delay 1 Sec
                            ;(call subroutine)
RJMP Again                  ;Goto to the beginning


;**************************
;Time delay 1 sec
;**************************
Delay_1sec:                 ;Delay subroutine
LDI R16,25                  ;Counter initialization
                            ;for the outer loop
Start1:                     ;Start point of outer loop
    LDI R17,100             ;Counter initialization for
                            ;the first inner loop
    Start2:                 ;Start point of the first inner loop
        LDI R18,100         ;Counter initialization for
                            ;the second inner loop
        Start3:             ;Start point of the first inner loop
            NOP             ;No operation for one Machine Cycle
            DEC R18         ;Counter decrement
        BRNE Start3         ;Check if the second inner loop
                            ;is completed
        DEC R17             ;Counter decrement
    BRNE Start2             ;Check if the first inner loop
                            ;is completed
    DEC R16                 ;Counter decrement
BRNE Start1                 ;Check if the outer loop
                            ;is completed
RET
```

Note
The total time delay is based on the main clock frequency of the microcontroller. That means that in the execution of programs that include time delays, time deviations can be occurred.

Application 5.3 – Successive LED activation with logical shift

Based on the circuit of figure 5.26, every bit of register PORTD affects directly the corresponding LED status. In this example, every next LED will be lit in turn (LED0 and LED1 to LED7). Figure 5.27 shows the LED states as well as the needed content of register PORTD.

LED status	PORTD	HEX	Operation
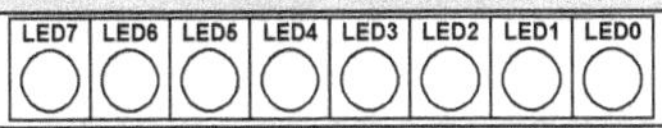 LED7–LED0: all off	00000000	00	Before load value
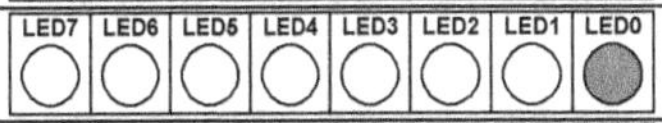 LED0 on	00000001	01	Load initial value
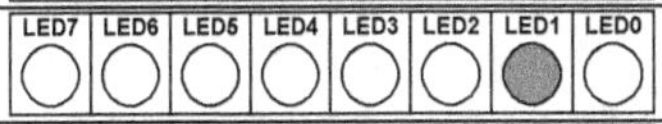 LED1 on	00000010	02	Logical shift left
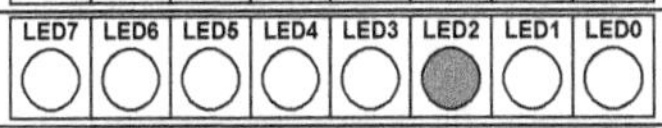 LED2 on	00000100	04	Logical shift left
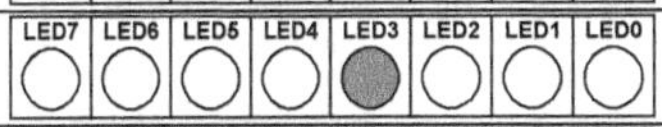 LED3 on	00001000	08	Logical shift left
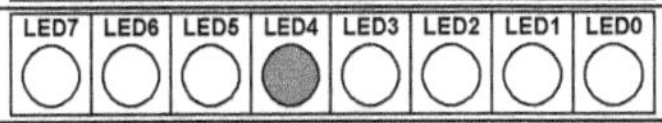 LED4 on	00010000	10	Logical shift left
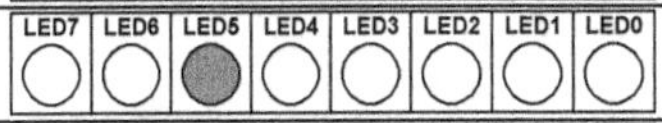 LED5 on	00100000	20	Logical shift left
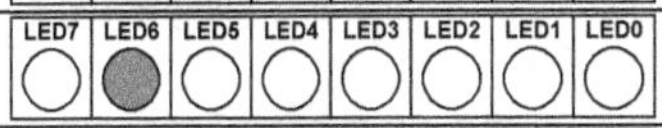 LED6 on	01000000	40	Logical shift left
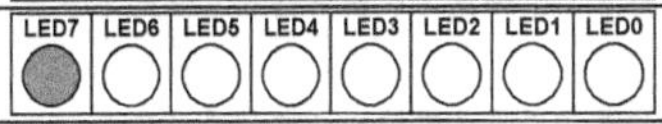 LED7 on	10000000	80	Logical shift left

Figure 5.27 Successive LED activation

The following code (code 5.6) performs the logical shifts with intermediate time delay. The flow chart diagram (fig. 5.28) shows the loop for performing the logical shift. In the same code, a time delay subroutine is used (as presented in the previous application).

Code 5.6

```
;Place here the proper INC file for your microcontroller model
;(if needed), e.g. ATmega32/ATmega32A => m32def.inc/m32Adef.inc,
;ATmega328 => m328def.inc

.INCLUDE "include/m32Adef.inc"

;**************************
;Stack initialization
;**************************
LDI  R16,HIGH(RAMEND)
OUT  SPH,R16
LDI  R16,LOW(RAMEND)
OUT  SPL,R16
```

```
.DEF LEDnum=R20
.DEF ShiftNUM=R22
.DEF on=R21

;**************************
;Main program
;**************************
LDI on,0xFF
OUT DDRD,on

init:
LDI LEDnum,1
LDI ShiftNUM,8

Again:
        OUT PORTD,LEDnum
        RCALL delay_1sec
        LSL LEDnum
        DEC ShiftNUM
BRNE Again
RJMP init


;**************************
;Time delay 1 sec
;**************************
delay_1sec:

LDI R16,25
Start1:
        LDI R17,100
        Start2:
                LDI R18,100
                Start3:
                        NOP
                        DEC R18
                BRNE Start3
                DEC R17
        BRNE Start2
        DEC R16
BRNE Start1
RET
```

As shown in the flow diagram (fig. 5.28), seven shifts are performed (the first LED is activated from the initial value of PORTD). Through the shift operation, the activation is "transferred" to the next LED. The `LEDnum` represents the LED under activation, and the `ShiftNUM` counts the number of shifts.

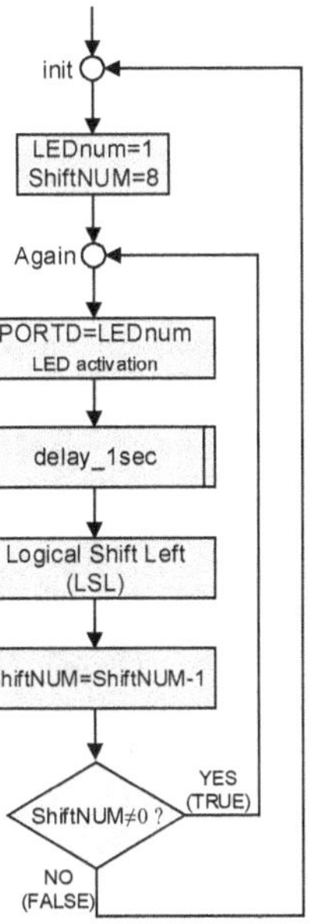

Figure 5.28 Flow chart diagram

Application 5.4 – Traffic light control

The traffic light control can be achieved by using LED control and time delay code. For the traffic light operation, three LEDs will be used (red, orange, green). The controlling procedure is based on LED activation and the corresponding delay (for every light). Figure 5.29 shows the operation phases of the traffic light.

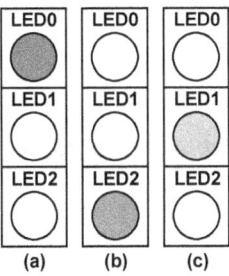

Figure 5.29 Traffic light phases

Every phase has a specific time duration. In current application, the corresponding phases are implemented as follows:

(a) Red light (10 seconds duration)
(b) Green light (5 seconds duration)
(c) Orange light (2 seconds duration)
(d) return to phase (a)

Figure 5.30 shows the circuit with the three connected LEDs to the microcontroller (LED0=red, LED1=orange, LED2=green).

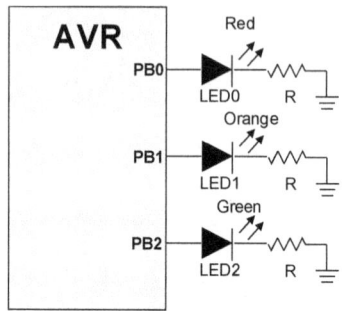

Figure 5.30 Traffic light circuit

For the circuit operation, the pins PB0, PB1 and PB2 of port B will be set as outputs. In the next step, the proper LED has to be activated with the corresponding time delay before deactivation and the next phase. The logic of the application operation is as follows:

```
Set pins PB0, PB1, PB2, as outputs
  AGAIN:
        Activation of LED0 (red)
        Delay 10 Sec
        Deactivation of LED0
        Activation of LED2 (green)
        Delay 5 Sec
        Deactivation of LED2
        Activation of LED1 (orange)
        Delay 2 Sec
        Deactivation of LED1
  Return to AGAIN
```

As shown above, the logic of operation is quite simple. On the other hand, the implementation of the time delay for 10, 5 and 2 seconds will be achieved by executing multiple times the subroutine of the 1 Sec delay (as used in previous applications). In many cases, different delays have to be implemented and thus, a new macroinstruction will be developed. The needed time delay will be set by the microinstruction parameter.

In other words, the programmer will decide how many seconds the time delay will be (as a multiplicand of the 1 Sec delay). The macroinstruction code is as follows:

Code 5.7

```
.MACRO delay_sec         ;Macroinstruction of adapted delay

LDI R19,@0               ;Number of iterations
                         ;for the 1 Sec delay
    again_sec:
```

```
        RCALL delay_1sec    ;Call of the delay subroutine
        DEC R19
  BRNE again_sec

.ENDMACRO
```

As shown in the above code, the register R19 is used as a counter within the loop which calls the delay subroutine of 1 Sec. The microinstruction is called from the main program with the number of delay in seconds as a parameter. For example, if the microinstruction is called by writing `delay_sec 5`, the value 5 will be loaded in register R19 and thus, the corresponding loop will be executed 5 times for a time delay of 5 seconds.

Note
The microinstruction can be replaced from a conventional subroutine (using the RCALL for the call). In such a case, the content of register R19 has to be initialized in order to set the duration of the delay (in seconds).

The complete application code is following.

Code 5.8

```
;Place here the proper INC file for your microcontroller model
;(if needed), e.g. ATmega32/ATmega32A => m32def.inc/m32Adef.inc,
;ATmega328 => m328def.inc

.INCLUDE "include/m32Adef.inc"

;********************************
;Declaration of constant values
;********************************
  .EQU red=0   ;Constant declaration for the red LED
  .EQU ora=1   ;Constant declaration for the orange LED
  .EQU gre=2   ;Constant declaration for the green LED

;**************************
;Delay macroinstruction
;**************************
.MACRO delay_sec            ;Macroinstruction of adapted delay

LDI R19,@0                  ;Number of iterations
                            ;for the 1 Sec delay
  again_sec:
        RCALL delay_1sec    ;Call of the delay subroutine
        DEC R19
  BRNE again_sec
.ENDMACRO

;**************************
;Stack initialization
;**************************
LDI R16,HIGH(RAMEND)
OUT SPH,R16
```

```
LDI R16,LOW(RAMEND)
OUT SPL,R16


;************************
;Set outputs
;************************
 SBI DDRB,0              ;Pin PB0=output
 SBI DDRB,1              ;Pin PB1=output
 SBI DDRB,2              ;Pin PB2=output


;************************
;Main program
;************************
 AGAIN:
       SBI PORTB,red     ;Red LED activation
       delay_sec 10      ;10 Sec delay
       CBI PORTB,red     ;Red LED deactivation
       SBI PORTB,gre     ;Green LED activation
       delay_sec 5       ;5 Sec delay
       CBI PORTB,gre     ;Green LED deactivation
       SBI PORTB,ora     ;Orange LED activation
       delay_sec 2       ;2 Sec delay
       CBI PORTB,ora     ;Orange LED deactivation
RJMP AGAIN


;************************
;Time delay 1 sec
;************************
delay_1sec:

LDI R16,25
Start1:
      LDI R17,100
      Start2:
            LDI R18,100
            Start3:
                  NOP
                  DEC R18
            BRNE Start3
            DEC R17
      BRNE Start2
      DEC R16
BRNE Start1

RET
```

Application 5.5 – Controlling a LED with a button

As a first step, a circuit where a LED is activated while the button (the switch has been replaced by a button) is pressed will be analyzed (fig. 5.31).

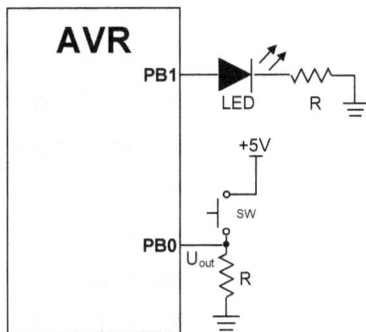

Figure 5.31 Controlling a LED with a button

The following program checks constantly the button state and activates the LED (while the button is pressed).

Code 5.9

```
;Place here the proper INC file for your microcontroller model
;(if needed), e.g. ATmega32/ATmega32A => m32def.inc/m32Adef.inc,
;ATmega328 => m328def.inc

.INCLUDE "include/m32Adef.inc"

.EQU Button=0          ;Button at Pin 0
.EQU LED=1             ;LED at Pin 1

;**************************
;Main program
;**************************
main:

    CBI DDRB,Button    ;Pin 0 of port B will be set as input
    SBI DDRB,LED       ;Pin 1 of port B will be set as output

scan:
        SBIS PINB,Button    ;If pin 0 (button) is activated, then
                            ;the next instruction is bypassed
        RJMP scan           ;Return for reading the button state
        SBI PORTB,LED       ;LED activation
        CBI PORTB,LED       ;LED deactivation
RJMP scan
```

Figure 5.32 shows the flow chart diagram for the above code.

Initially, the data direction of the pins PB0 and PB1 must be set using the instructions CBI (Clear BIT) and SBI (Set BIT) respectively. Using these instructions, only the two pins are affected. The instruction OUT DDRB,R16 (with R16=0x00 or 0xFF) is not used due to the fact that affects all the port pins.

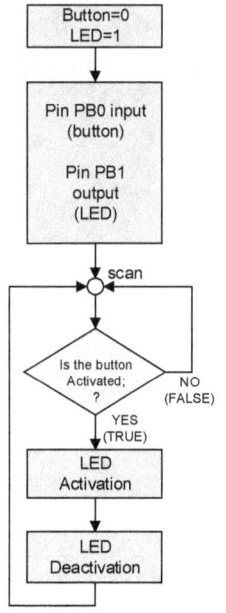

Of course, in such a case, a different port could be used for different data direction, but the current program supports only the above operation (the button is checked constantly for LED activation).

The SBIS instruction checks the input pin PB0. If this pin is activated (PB0=1), then the next instruction is not executed (the flow does not return to button check) and the LED is activated and deactivated for a moment. Using this approach, the LED is only activated while the button is pressed. After the above LED operation, the execution flow returns to the button check.

Figure 5.32 Flow chart

The execution flow returns also to button check while PB0=0 (checking with the SBIS instruction). Figure 5.33 shows the execution flow based on the button state.

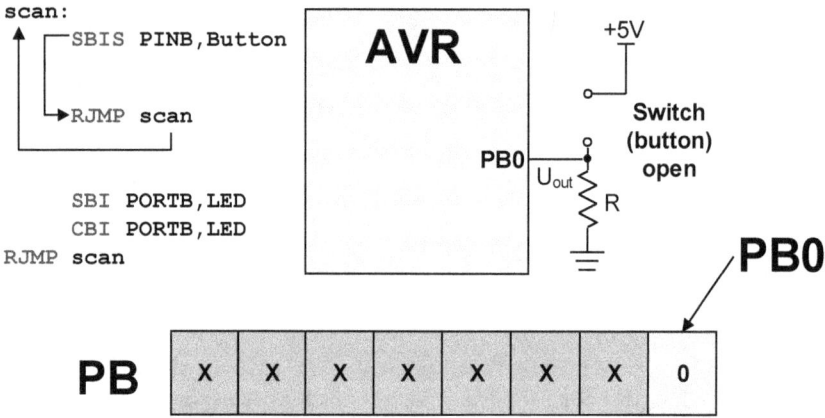

Figure 5.33a Change execution flow based on pin status

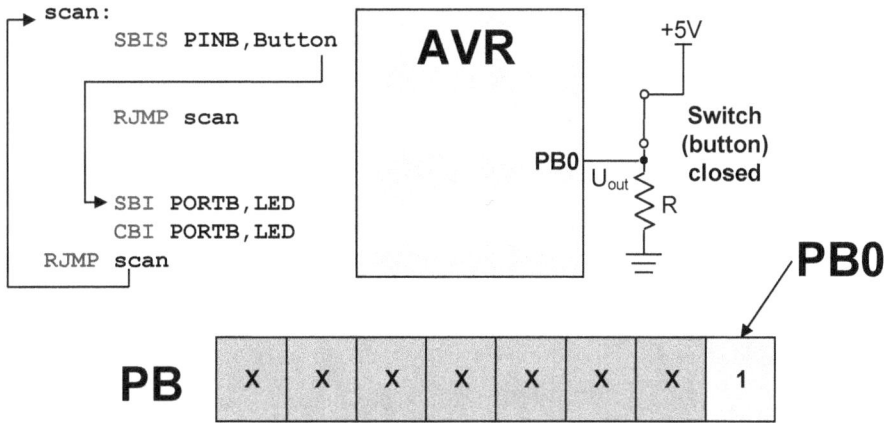

Figure 5.33b Change execution flow based on pin status

Another implementation is to invert the LED state every time the button is pressed (toggle operation). Like the above case, in the toggle operation, the button state will be checked constantly. With a special condition check the previous button state will be checked (if the current button state is HIGH and the previous state is LOW).

Additionally, an important issue in this application is the mechanical construction of the button which causes multiple activations with only one press. More precisely, a single button press is detected as multiple due to the mechanical oscillation of the button. Using an oscillator, the above mechanical oscillation can be measured. Figure 5.34 shows the oscillation when the button is pressed. On the other hand, figure 5.35 shows the oscillation when the button is released.

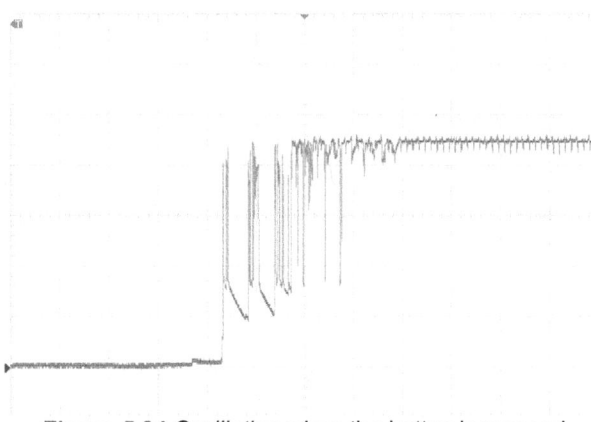

Figure 5.34 Oscillation when the button is pressed

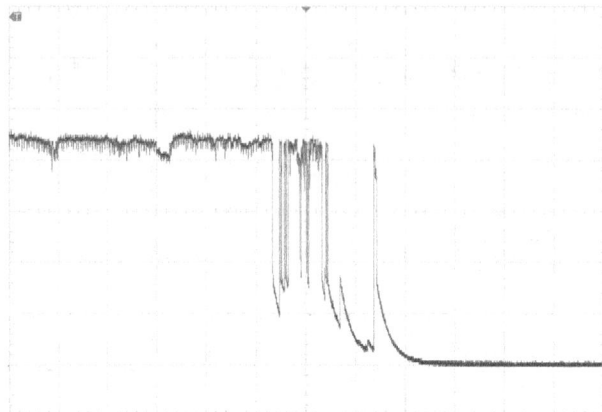

Figure 5.35 Oscillation when the button is released

In order to avoid this phenomenon, a time delay of few dozen of milliseconds is enough until the button calm. Another solution is to use a capacitor in parallel with the button.

Application 5.6 – Button circuit simplification using the internal Pull-Up resistor

As mentioned in previous section, the microcontroller has internal Pull-Up resistors that are connected internally to the pins of the I/O ports. The activation is achieved by writing HIGH to the desired pin, while has been set as input. With this approach, the switch-button circuit is simplified (figure 5.36). For verifying the corresponding operation in this application, a LED will be activated every time the button is pressed.

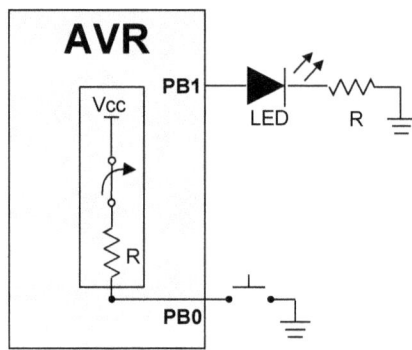

Figure 5.36 Simplified switch circuit

The whole code of the application follows.

Code 5.10

```
;Place here the proper INC file for your microcontroller model
;(if needed), e.g. ATmega32/ATmega32A => m32def.inc/m32Adef.inc,
;ATmega328 => m328def.inc
```

```
.INCLUDE "include/m32Adef.inc"

.EQU Button=0          ;Button at pin 0
.EQU LED=1             ;LED at pin 1

;***************************
;Main program
;***************************

main:

    SBI PORTB,Button   ;Pull Up resistor activation
                       ;at pin PB0
    SBI DDRB,LED       ;The pin 1 of port B
                       ;will be set as output


scan:
    SBIC PINB,Button   ;If the pin 0 (button)
                       ;is activated
                       ;(PB0=LOW=0V),
                       ;then the next instruction is bypassed

    RJMP scan          ;Return to check button

    SBI PORTB,LED      ;LED activation
    CBI PORTB,LED      ;LED deactivation
RJMP scan
```

In the above code that uses the internal Pull Up resistor of the pin PB0, there are two main points that differ as compared to the previous applications.

(a) Pull Up resistor activation

The activation is achieved by writing 1 to pin 0 (connected button), despite the fact that the pin 0 is set as input. The `SBI PORTB,Button` instruction writes the value 1 in an input pin!. The default data direction of port B is input because the initial value of register `DDRB` is zero. The same result is generated if the `CBI DDRB,Button` is used before the SBI instruction.

(b) Checking pin for a LOW level

The `SBIC` instruction bypasses the next instruction (which returns the execution flow to button check), if a LOW level has been detected. This instruction performs the inverse operation of the SBIS.

Application 5.7 – Direct LED control using buttons (switches)

Let assume that there is a microcontroller circuit with 8 LEDs and 8 switches respectively. Figure 5.37 shows the electrical connections of input and output circuits. Of course, the switch and LED circuit can be implemented differently using the previously mentioned methods. For example, the LEDs can be connected with the positive pins to 5 volts. Thus, every LED will be lit using a LOW signal (changes also have to be implemented in the code). The program of the current application "transfers" directly the switch states to the LEDs. This is achieved by "copying" the switch states (PIND) directly to the pins of port B (PORTB).

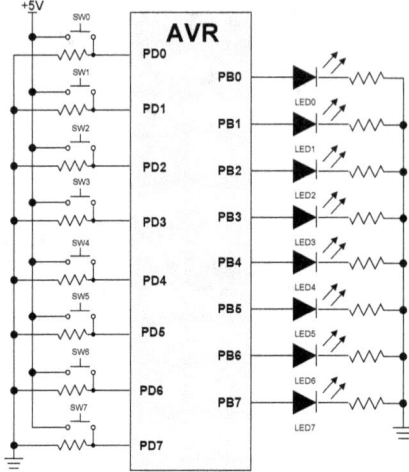

Figure 5.37 Controlling 8 LEDs with switches

As shown in figure 5.37, the inputs are connected to Pull Down circuits. Thus, when the button is pressed (switch activation), 5 volts appear on the corresponding pin. The above operation is supported by the following code:

Code 5.11

```
;Place here the proper INC file for your microcontroller model
;(if needed), e.g. ATmega32/ATmega32A => m32def.inc/m32Adef.inc,
;ATmega328 => m328def.inc

.INCLUDE "include/m32Adef.inc"

.EQU input=0x00             ;Value for setting inputs
.EQU output=0xFF            ;Value for setting outputs

LDI R16,input               ;Load R16 for setting inputs

LDI R17,output              ;Load R17 for setting outputs


OUT DDRD,R16                ;PD0 to PD7, inputs (switches)
OUT DDRB,R17                ;PB0 to PB7, outputs (LEDs)

again:
        IN  R16,PIND        ;Read switches states
        OUT PORTB,R16       ;Update LEDs states based on
                            ;switches states

RJMP again                  ;Repeat operation forever
```

Figure 5.38 shows as an example how the switches states is "transferred" to LEDs of port B through the registers PIND, R16 and PORTB.

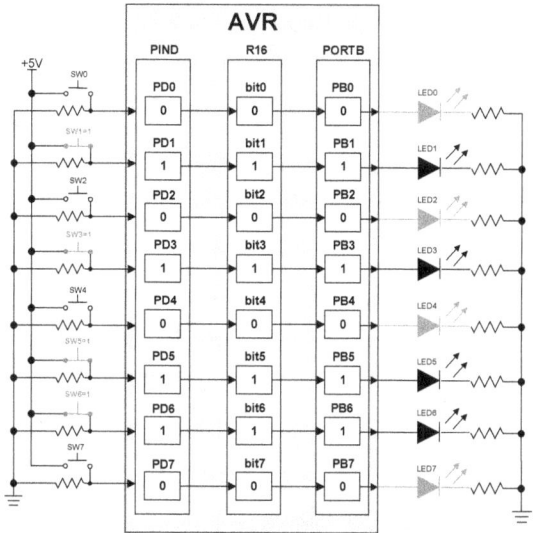

Figure 5.38 "Transferring" the switches states to LEDs

Figure 5.38 shows that the switches SW1, SW3, SW5 and SW6 are activated. Thus, the 5V signal (logical 1) is applied in the corresponding inputs. Based on that inputs, the content of register PIND (from the most significant bit to the least significant bit) will be 01101010. This content is loaded in R16 and then in register PORTB where the LEDs are connected to the corresponding pins.

Note
From this point forward, the lab activities start.

Useful information about lab activities
1. For the microcontroller operation, a power supply is needed as well as an external reset circuit for initializing.

2. In some cases, the power supply is supported by the circuit for the physical programming of the microcontroller (e.g. USBasp).

3. The chapter applications can be also used as lab exercises.

4. Initially, the implementation of the chapter applications can be asked by the professor before the lab exercise study and development.

LABORATORY EXERCISE 1
Basic electrical circuits

GOAL
The goal of this lab exercise is the implementation of electrical circuits that consists of resistors and diodes. Additionally, real measurements will be made by students in order to understand the meaning of current, voltage divider and the diode operation. The power supply can be provided by a single battery or by a stabilized laboratory power supply of low current. The measurements will be listed in a table and conclusions will be extracted.

Step 1 (Recognize and measure resistors)
Every resistor has a color code which represents the value in ohms ($\Omega\mu$). This code represents also the value tolerance. There are resistors with 4, 5 or 6 color zones. In this exercise, resistors with 4 and 5 color zones will be used (see the following figure).

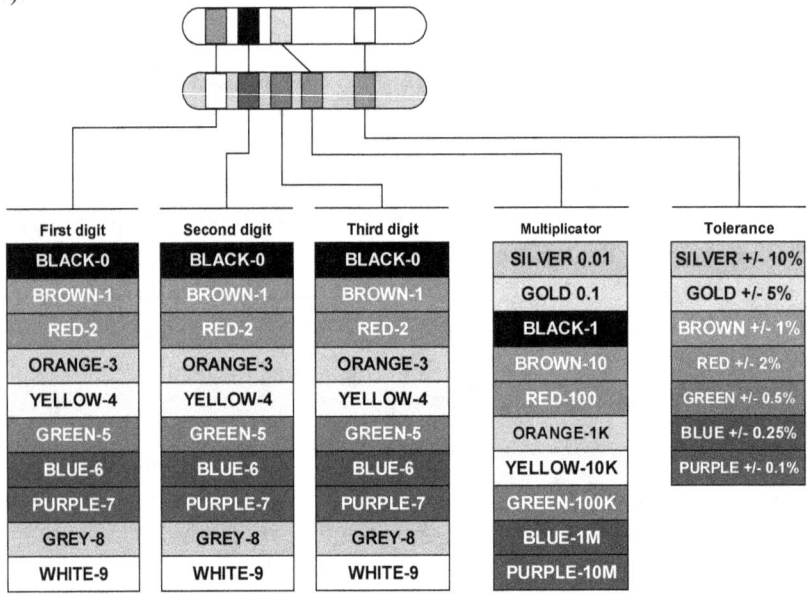

The physical form of four resistors (¼ W) is shown in the next figure.

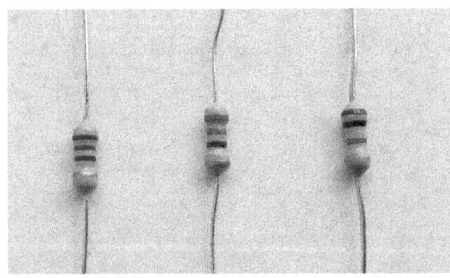

Recognize the resistors 150Ω, 220Ω, 470Ω, 680Ω, 1KΩ, 2.2KΩ in the Lab and fill the following table:

Resistor	Color code	Tolerance	Measurement with ohmometer	Is the measurement in the tolerance range? (YES/NO)
150Ω				
220Ω				
470Ω				
680Ω				
1KΩ				
2.2KΩ				

Step 2 (Voltage and current measurement)
For fast and easiest circuit connections, a breadboard will be used.

As shown in the above figure, the breadboard has a number of available holes (tie points) where the pins of electronic elements can be placed. These holes, are connected internally within the under layer of the breadboard. The next figure shows the internal connections.

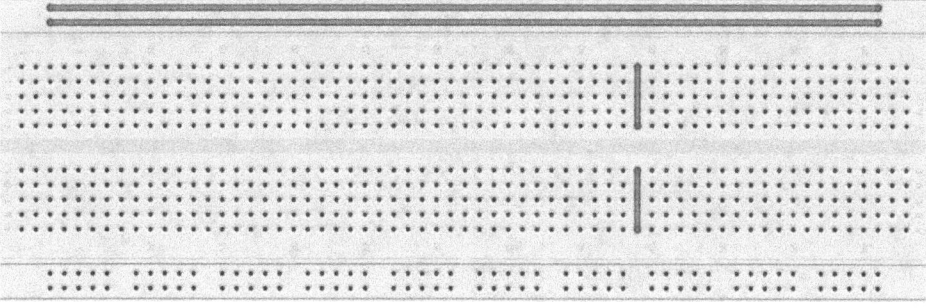

The columns are connected vertically and belong to two electrically separated areas. There are also horizontal connections in the upper and bottom areas of the breadboard. In these lines, the power supply (e.g. 5V) and the ground (GND) are connected from the corresponding circuit for giving multiple signal tie points availability.

Place the four resistors according to the following figure:

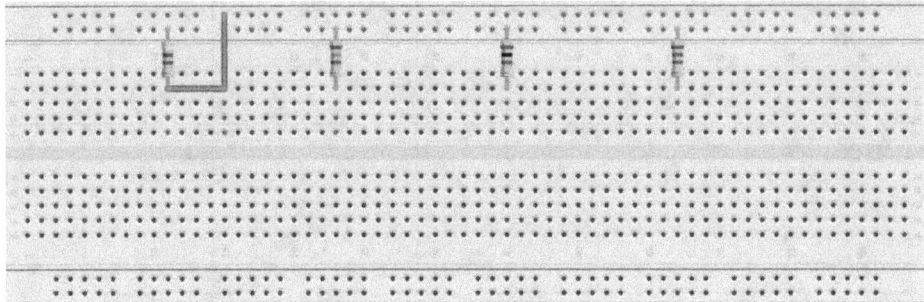

Note that, the first pin of the resistor is connected in the upper red line where the +5V (or other voltage) will be applied. The other pin will be connected in the ground (GND) line. Connect the rest of the resistors like the first one (on the left).

Step 3
Recognize the resistors and fill the following table:

Resistor (from the left)	Nominal value	Measured value
1st		
2nd		
3rd		
4th		

Step 4
Apply external power supply according to your professor's guide lines (+U in the upper red line and – in the upper blue line) and make current measurements. The following figure shows how to measure the circuit current using an ampere meter (or a multimeter).

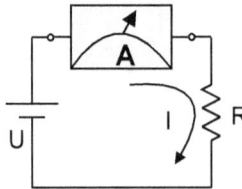

Measure the current and fill the following table:

Experimental measurements		
U = ... V		
R (resistor)	I (current)	Theoretical calculation of current
470Ω		
680Ω		
1KΩ		
2.2KΩ		

Step 5
Draw in the following scaled paper the current amplitude as a function of the resistor.

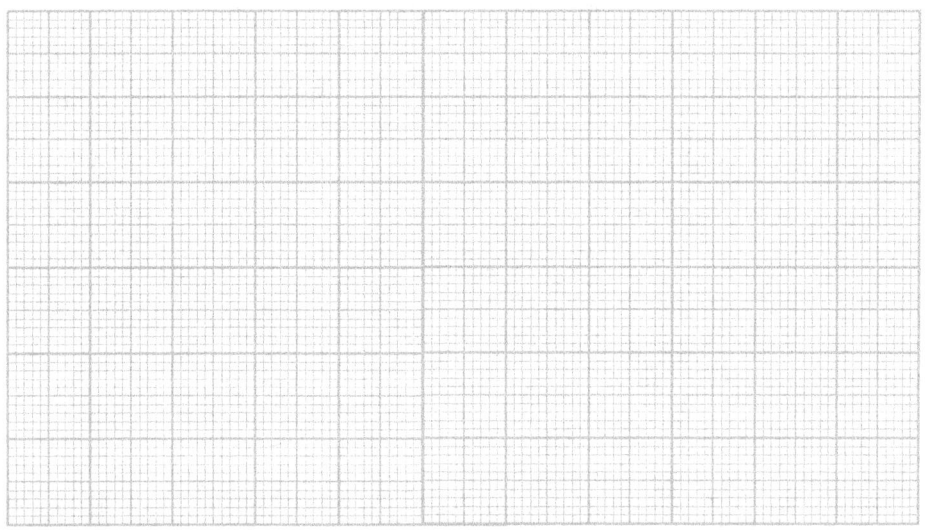

Step 6
Implement the following circuit by changing the resistor values and fill the table.

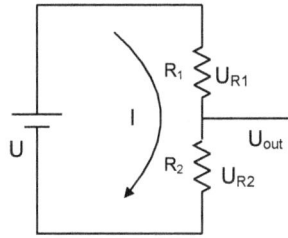

Experimental measurements

U = ... V

R1	R2	U$_{R1}$	U$_{R2}$	U$_{out}$	I
470Ω	470Ω				
680Ω	680Ω				
1KΩ	1KΩ				
470Ω	680Ω				
470Ω	1KΩ				
470Ω	2.2KΩ				
680Ω	470Ω				
1KΩ	470Ω				
2.2KΩ	470Ω				

Step 7
Make the theoretical calculations of the above circuit and fill the following table:

Theoretical calculations

U = ... V

R1	R2	U$_{R1}$	U$_{R2}$	U$_{out}$	I
470Ω	470Ω				
680Ω	680Ω				
1KΩ	1KΩ				
470Ω	680Ω				
470Ω	1KΩ				
470Ω	2.2KΩ				
680Ω	470Ω				
1KΩ	470Ω				
2.2KΩ	470Ω				

Comment the circuit operation. Focus your attention on the voltages U and U$_{out}$.

Step 8
Implement the following circuit by changing resistors and diodes and fill the measurement table.

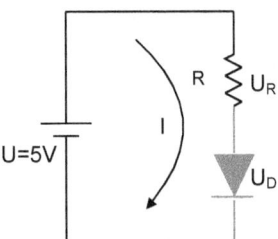

Experimental measurements

U = ... V

R	Red LED U$_D$		I

R		
150Ω		
220Ω		
470Ω		
	Green LED	
R	U_D	I
150Ω		
220Ω		
470Ω		
	Yellow LED	
R	U_D	I
150Ω		
220Ω		
470Ω		
	Blue LED	
R	U_D	I
150Ω		
220Ω		
470Ω		

Step 9
Draw in the scaled paper the voltage U_D according to resistor and diode values combinations.

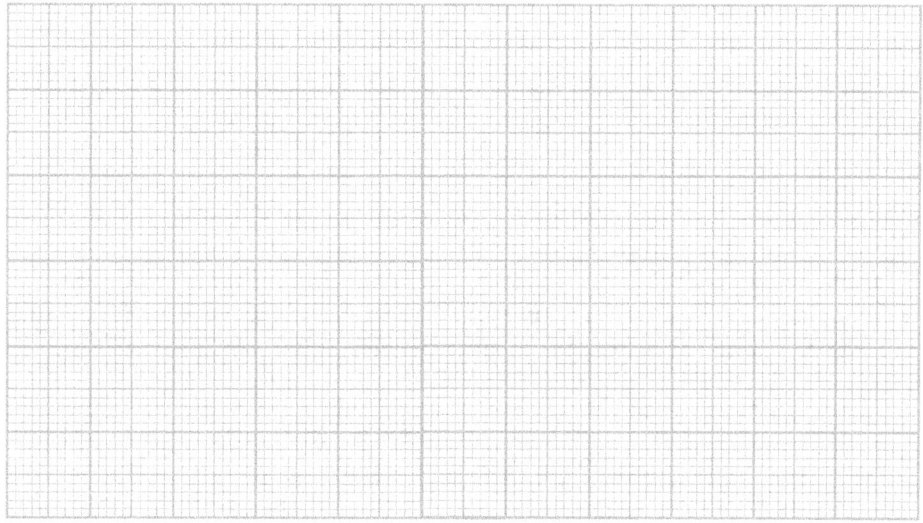

Step 10 (homework)
Use a constant resistor and a 5KΩ potentiometer in order to control the circuit current where also a diode is connected. Measure the diode current and voltage while rotating the potentiometer (try diode voltages 0.3V, 0.4V, etc). In this exercise you will try to sketch the characteristic diode diagram (I-V) based on the experimental measurements.

LABORATORY EXERCISE 2
Time Delay

GOAL
In this exercise a time delay will be developed and measured for controlling a LED.

Step 1
Develop a program for inverting the LED status (on/off/on, etc) every 0.5 seconds. The LED is connected at PD0 (pin 0 of port D). For the program operation a circuit has to be implemented (see next figure). Initially, observe the program operation on the LED state.

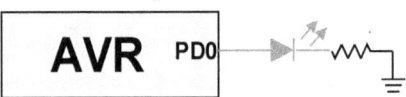

Step 2
Fill the following signal attributes based on the expected measurements on pin PD0.

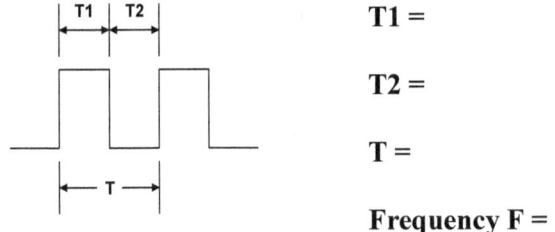

T1 =

T2 =

T =

Frequency F =

Step 3
Connect one channel of the oscilloscope as described in the next figure. Set properly the time base (Time/Div) and the voltage scale (Volts/Div) on the oscilloscope in order to display the measured signal at the right size.

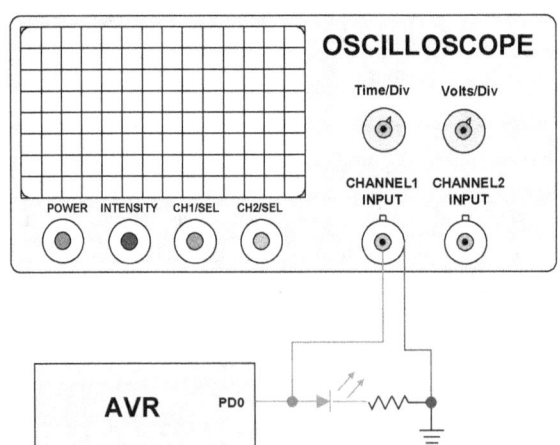

Step 4
Draw in the scaled paper the waveform which is displayed on the oscilloscope screen.

Step 5
Compare and comment the previous measurements in relation with the expected ones and fill a corresponding table.

Step 6
Study the following code:

```
Delay:
LDI R16,25
Start1:
        LDI R17,25
        Start2:
           LDI R18,7
           Start3:
                NOP
                NOP
                NOP
                NOP
                DEC R18
            BRNE Start3
               DEC R17
        BRNE Start2
        DEC R16
BRNE Start1
```

(a) Calculate theoretically the code execution duration for clock frequency of 1MHz and 16MHz

(b) Test the above code by using physical ports of the microcontroller and measure the execution time using an oscilloscope

LABORATORY EXERCISE 3
Switch circuits

GOAL
The goal of this lab exercise is to study and use switch circuits for sending signals to the microcontroller. Using such switches, a user may enter settings to an application.

Step 1
Study the circuit of the next figure and answer to the following questions.

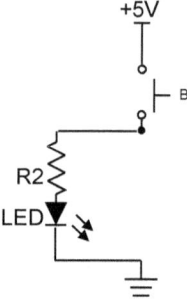

(1) Write down how the circuit current will be calculated
(2) If the LED represents the input signal to the microcontroller, categorize the circuit as Pull Up or Pull Down
(3) Draw the equivalent circuit when the switch is closed (B, button)

Step 2
Modify the above circuit in order to turn off the LED when the switch is closed (B, button).

Step 3
Implement a real circuit based on the following block diagram. Select a suitable resistor for the LEDs in order to allow a current flow of 15 to 25mA.

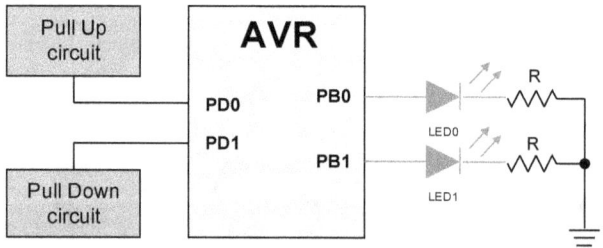

Step 4
Develop a program in order to activate the LEDs LED0 and LED1 by detecting a state change in the corresponding Pull Up and Pull Down circuits.

Step 5
Test the following circuit

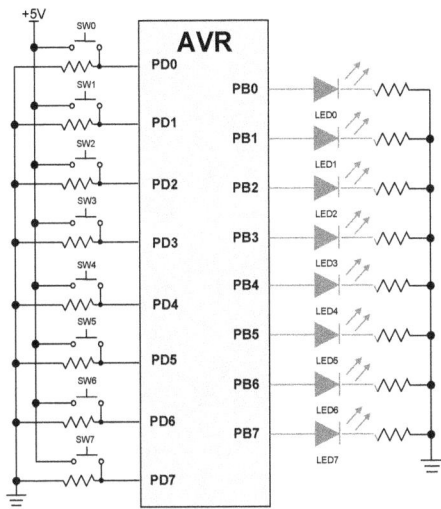

Step 6
Develop a program for supporting the following operations:

(a) Read the input states at the pins PD0 to PD7
(b) Activate the 8 LEDs based on the states of the pins PD0 to PD7
(c) Perform four logical shifts to the left and four logical shifts to the right with an intermediate delay of 1 Sec

6 Display units manipulation

Content-Goals
Display units are important devices for displaying useful information for the end-user and also important data for the developer. In this chapter the most popular display devices such as the seven segment unit and LCD screen will be programmed.

Chapter Contents
6.1 Seven segment display manipulation
6.2 LCD screen

6.1 Seven segment display manipulation

LEDs constitute the simplest display elements but are impractical for displaying more complex data. On the other hand, for displaying binary data, LEDs constitute the most suitable solution due to the fact that the LED state represents directly the binary value (1=LED on, 0=LED off). Moreover, LEDs have two significant disadvantages:

(a) the binary representation is impractical for large numbers (e.g. with many bits)

(b) users are more familiar with decimal numbers and thus, many LEDs and more complex circuits are needed for such a representation

For displaying numbers in decimal form, one or more Seven Segment Displays (SSDs) are used. The SSD is a unit (or device) which consists of eight LEDs (7 LEDs for displaying digits and one LED for the decimal point). These LEDs are embedded in the SSD unit and can be activated separately like a classical LED using a resistor and forward biasing.

Figure 6.1a shows the form of the SSD unit. Every segment consists of a LED which is lit when it is forward biased (+ to the anode, - to the cathode). The internal connection of the SSD unit is shown in figure 6.1b. The SSD unit of figure 6.1b, is a common cathode unit where all the LED cathodes are connected together and are used for the ground. On the other hand, every segment has a separate positive pin (LED anode).

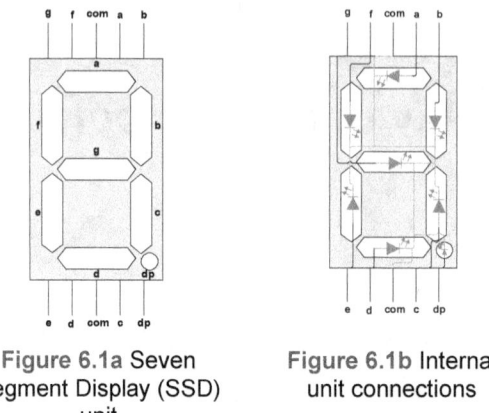

Figure 6.1a Seven Segment Display (SSD) unit

Figure 6.1b Internal unit connections

Figures 6.1c and 6.1d show the physical from (both sides) of the SSD.

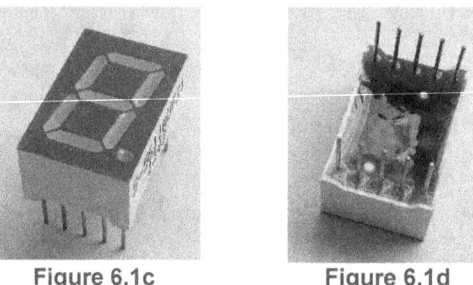

Figure 6.1c
Front view of the SSD

Figure 6.1d
Rear view of the SSD

Figure 6.1e shows an equivalent circuit of figure 6.1b.

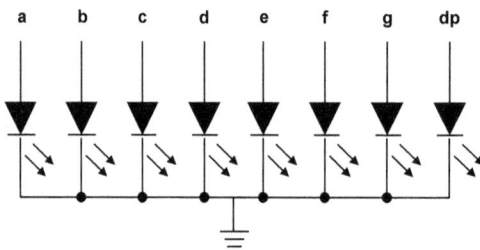

Figure 6.1e Internal diode circuit of the SSD (common cathode)

The next figure (figure 6.2) shows the two types of the SSD unit (CC-Common Cathode, CA-Common Anode). The only difference is how the LEDs are organized internally. Figure 6.2 shows also how all the LEDs of the two SSD units can be lit. In first SSD (CC type), all the LEDs are connected to a common ground. For lighting each segment (LED) a separate resistor must be used. In the Common Anode (CA) SSD, all the LED positive pins are connected to a common SSD pin and thus, for lighting each segment a LOW (0V) signal is required.

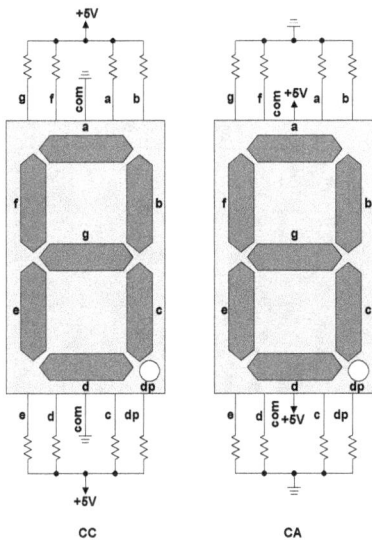

Figure 6.2 Lighting up all the LEDs
Common Cathode (CC) and Common Anode (CA)

Application 6.1 – Testing the SSD unit operation

In the first example, the SSD unit operation will be tested by lighting up each segment (a to g) in turn using a time delay of 1 second. Figure 6.3 shows the activation sequence (phases) of the SSD unit.

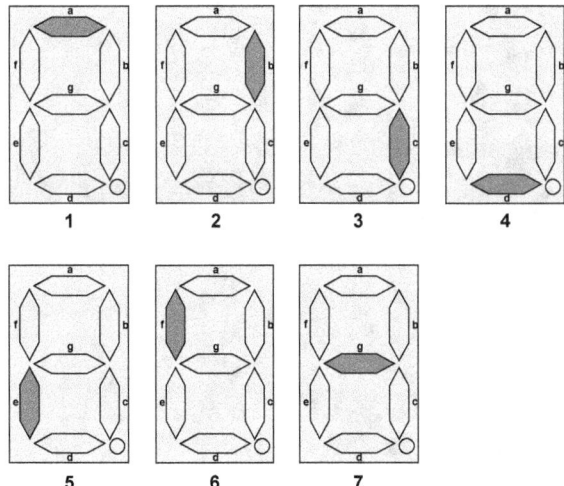

Figure 6.3 Activation sequence of the SSD unit

For the above implementation, a common cathode SSD unit will be used (fig. 6.4). The pins PB0 to PB6 (7 bits) of port B will be used for activating the seven segments of the SSD unit (the decimal point LED will not be used). For simplicity reasons the whole register will be used (PB7 is unconnected). In the practical circuit, 150-220Ω resistors will be used.

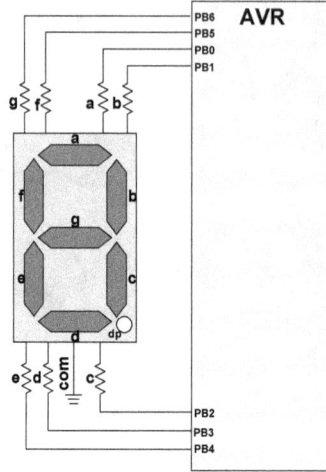

Figure 6.4 Control circuit for a SSD unit

The following code is for testing the SSD unit.

Code 6.1

```
;Place here the proper INC file for your microcontroller model
;(if needed), e.g. ATmega32/ATmega32A => m32def.inc/m32Adef.inc,
;ATmega328 => m328def.inc

.INCLUDE "include/m32Adef.inc"

;***************************
;Stack initialization
;***************************
LDI  R16,HIGH(RAMEND)
OUT  SPH,R16
LDI  R16,LOW(RAMEND)
OUT  SPL,R16
LDI  R20,0xFF          ;Load 11111111
                       ;for defining output
OUT  DDRB,R20          ;Set all the pins of port B as outputs
LDI  R22,0x00          ;Load 00000000 for LED deactivation

;***************************
;Main program
;***************************
start:
LDI  R21,7             ;7 iterations (one for each segment)
LDI  R23,0x01          ;Initially, the pin PB0 will be activated
                       ;(segment a)
again:                 ;Loop return point
OUT  PORTB,R23         ;Segment activation based on R23 content
ROL  R23               ;Left shift(prepare next segment)
RCALL delay_1sec       ;Time delay (1 second)
OUT  PORTB,R22         ;Deactivate all segments
```

```
DEC R21                 ;Counter decrement
BRNE again              ;While there are still segments for activation
                        ;return to 'again'
RJMP start              ;Repeat forever

;**************************
;Time delay 1 sec
;**************************
delay_1sec:             ;Delay subroutine

LDI R16,25
Start1:
    LDI R17,100
    Start2:
        LDI R18,100
        Start3:
            NOP
            DEC R18
        BRNE Start3
        DEC R17
    BRNE Start2
    DEC R16
BRNE Start1

RET
```

Segment operation control using a macroinstruction

Due to assembly code complexity (difficult organization and development) it is useful to include one or more code sections as a microinstruction in an external file. Thus, the time delay code will be embedded in a macroinstruction. The macroinstruction will be used inside code every time is needed.

Figure 6.5 shows the inclusion method of the external file as well as the usage method of the time delay macroinstruction.

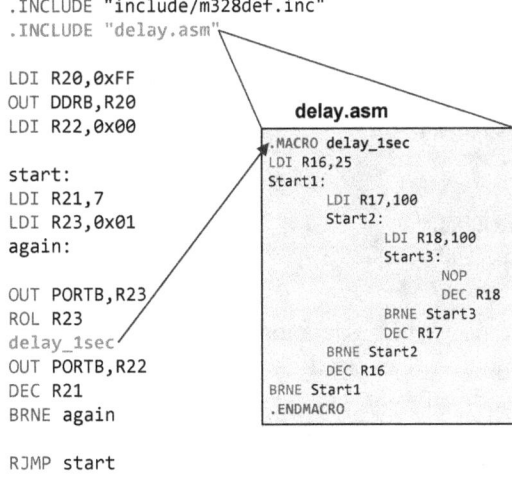

Figure 6.5 Using the macroinstruction (externally)

Application 6.2 – Displaying the numbers 0 to 9

For displaying the digits of the decimal numeric system, the digits 0 to 9 have to be displayed. Every SSD unit pin is connected to a pin of the port B. Thus, the port B contents affect directly the corresponding segment activation. Table 6.1 shows the register content for displaying the digits.

Table 6.1 Displaying the decimal system digits

SSD digit	Active segments	Register content (Output port) PB7 PB6 PB5 PB4 PB3 PB2 PB1 PB0 (seg: g f e d c b a)	HEX
0	a,b,c,d,e,f	0 0 1 1 1 1 1 1	0x3F
1	b,c	0 0 0 0 0 1 1 0	0x06
2	a,b,d,e,g	0 1 0 1 1 0 1 1	0x5B
3	a,b,c,d,g	0 1 0 0 1 1 1 1	0x4F
4	b,c,f,g	0 1 1 0 0 1 1 0	0x66
5	a,c,d,f,g	0 1 1 0 1 1 0 1	0x6D
6	a,c,d,e,f,g	0 1 1 1 1 1 0 1	0x7D
7	a,b,c	0 0 0 0 0 1 1 1	0x07
8	a,b,c,d,e,f,g	0 1 1 1 1 1 1 1	0x7F
9	a,b,c,d,f,g	0 1 1 0 1 1 1 1	0x6F

In this application, a loop will be developed for displaying the digits 0 to 9 using a time delay of 1 second before every digit display. As shown in table 6.1, every digit is displayed by activating the proper pins of the corresponding port (port B in the current application). Due to the fact that the decimal digits have to be available for reuse in any application, it is practical to store them in the program memory. The following code, displays the digits 0 to 9 on the SSD unit.

Code 6.2

```
;Place here the proper INC file for your microcontroller model
;(if needed), e.g. ATmega32/ATmega32A => m32def.inc/m32Adef.inc,
;ATmega328 => m328def.inc

.INCLUDE "include/m32Adef.inc"

;**************************
;Stack initialization
;**************************
LDI  R16,HIGH(RAMEND)
OUT  SPH,R16
LDI  R16,LOW(RAMEND)
OUT  SPL,R16
                            ;Set data Direction of port B
LDI  R18,0xFF
OUT  DDRB,R18

;**************************
;Main program
;**************************

begin:                      ;Load the initial address
                            ;of table (array) in the register
                            ;Z (16bit)
  LDI    ZL,LOW(digits*2)
  LDI    ZH,HIGH(digits*2)

                            ;Initialize loop
                            ;for displaying the 10digits
  LDI    R25,0

again:
  LPM                       ;Load R0 with the content of the
                            ;address which is stored
                            ;in register Z
  OUT    PORTB,R0           ;Write in portB
                            ;Display (bit values) from
                            ;table(program memory)
  RCALL  delay_1sec         ;1 second delay
  ADIW   Z,1                ;Increment of Z
                            ;for reading the next table element
  INC    R25                ;Increment of loop counter
  CPI    R25,10             ;Compare counter content with 10
  BRLT   again              ;While the counter content is less
                            ;than 10,(≤ 9), return to 'again'
  RJMP   begin              ;The whole program is executed
                            ;permanently

;**************************
;Time delay 1 sec
;**************************
delay_1sec:
LDI R16,25
```

```
Start1:
        LDI R17,100
        Start2:
                LDI R18,100
                Start3:
                        NOP
                        DEC R18
                BRNE Start3
                DEC R17
        BRNE Start2
        DEC R16
BRNE Start1
RET

;bit values for every digit (0 to 9)
digits: .DB 0x3F,0x06,0x5B,0x4F,0x66,0x6D,0x7D,0x07,0x7F,0x6F
```

Application 6.3 – Controlling the SSD unit by using multiplexing

In many cases there is the need to display multiple digits (e.g. displaying a temperature). For achieving this, multiple SSD units have to be used. In such a case, two critical issues have to be faced effectively:

(a) The required pins for lighting up so many LEDs (e.g. for three SSD units, 21 pins are required or 24 pins if the decimal point of the SSD units is also used) are usually more than the available pins that a microcontroller has. It must be noticed also that the application developer has also the goal to use as few pins as possible.

(b) The required current for lighting up so many LEDs (e.g. for three SSD units there are 21 LEDs and 24 LEDs if the decimal point of the SSD units is also used) cannot be supplied by the microcontroller

Due to current limitation, usually only one SSD unit can be active every time. On the other hand, for displaying more digits, more SSD units have to be used (this is not acceptable due to the mentioned limitation).

For solving the above issues, a special technique which is based on the human eye features will be used. If for example, the SSD unit is turned off for time slices less than 50mSec, then the human eye still read the displayed number. Thus, it is not necessary to keep activated all the SSD units together. As an example, assume that the number 15 will be displayed on two SSD units. The following pseudocode shows the logic behind the above technique.

```
Turn off the two SSD units
START
        Display the number 1 on the left SSD unit (duration<50mSec)
        Turn off the left SSD unit
        Display the number 5 on the right SSD unit (duration<50mSec)
        Turn off the right SSD unit
RETURN TO START
```

As shown in the above logic, only one SSD unit is active at any time, but the human eye reads always the whole number (number 15). The circuit for controlling the two SSD units is shown in figure 6.6. In this circuit, the pins of the SSD units are connected in parallel (the segment 'a' of the first SSD unit with the segment 'a' of the second SSD unit, etc).

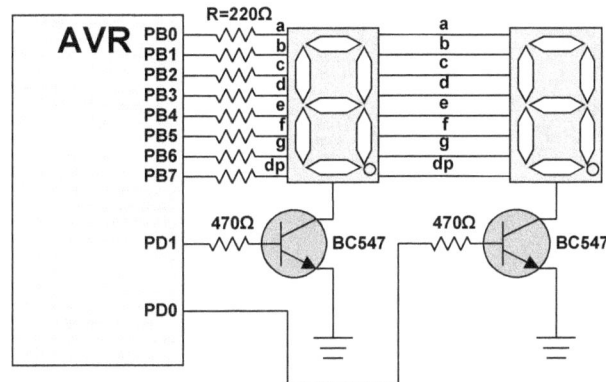

Figure 6.6 Control circuit of the two SSD units

The PB0 to PB7 signals are applied in the corresponding SSD unit segments through the 220Ω resistors. This is not enough to light up the SSD units because the common SSD pin is not connected to ground. For controlling each SSD unit separately, a programmable switch has to be installed in every common pin. Each programmable switch will be implemented with a transistor which will be controlled from a microcontroller pin. The transistor switch will be active when a suitable current is flown in its base through a 470Ω resistor (when the corresponding microcontroller pin is HIGH). Table 6.2 shows the necessary signals for displaying the number 1 on the left SSD unit and then on the right SSD unit.

Table 6.2 Activation example (SSD units)

PB 0 a	PB 1 b	PB 2 c	PB 3 d	PB 4 e	PB 5 f	PB 6 g	PB 7 dp	PD 0	PD 1	SSD unit	Display number
0	1	1	0	0	0	0	0	0	1	Left	1
0	1	1	0	0	0	0	0	1	0	Right	1

Figure 6.7 shows the active 'lines' in order to display the number 1 on the left SSD unit.

196 CHAPTER 6

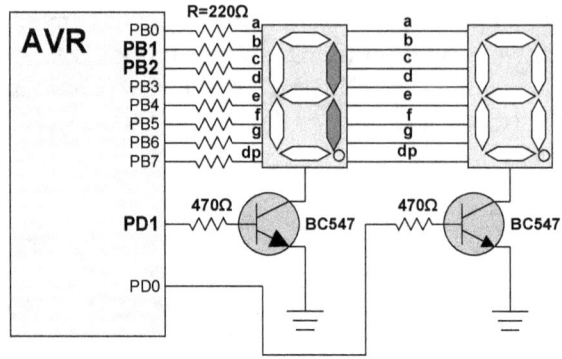

Figure 6.7 Displaying '1' on the left SSD

The corresponding code follows.

Code 6.3

```
;Place here the proper INC file for your microcontroller model
;(if needed), e.g. ATmega32/ATmega32A => m32def.inc/m32Adef.inc,
;ATmega328 => m328def.inc

.INCLUDE "include/m32Adef.inc"

;*************************
;Stack initialization
;*************************
LDI  R16,HIGH(RAMEND)
OUT  SPH,R16
LDI  R16,LOW(RAMEND)
OUT  SPL,R16

LDI  R18,0xFF
OUT  DDRB,R18

SBI  DDRD,0
SBI  DDRD,1

;*************************
;Display Macroinstruction
;*************************
.MACRO printdigit

SBI PORTD,@0

    LDI    ZL,LOW(digits*2)
    LDI    ZH,HIGH(digits*2)

    ADD    ZL,R19          ;Position select (digit)

    LPM                    ;Read from program memory
    OUT    PORTB,R0        ;Display digit
```

```
.ENDMACRO

;***************************
;Main program
;***************************

restart:

;***************************
;Display digits 0-9
;***************************

LDI R20,0
loop:
      MOV R19,R20
      printdigit 0
      RCALL delay_1sec
      INC R20
      CPI R20,10
BRLT loop

RJMP restart

;***************************
;Time delay 1 sec
;***************************
delay_1sec:
LDI R16,25
Start1:
      LDI R17,100
      Start2:
            LDI R18,100
            Start3:
                  NOP
                  DEC R18
            BRNE Start3
            DEC R17
      BRNE Start2
      DEC R16
BRNE Start1
RET

digits:  .DB 0x3F,0x06,0x5B,0x4F,0x66,0x6D,0x7D,0x07,0x7F,0x6F
```

Application 6.4 – Display four digits on multiple SSD units

Based on the above display circuits, a new circuit with four SSD units will be developed. The new SSD control circuit is based on repeated elements of the above circuits (fig. 6.8a). Based on the same approach, more SSD units can be used. In current application, an additional common anode SSD unit is also proposed.

Note
The SSD unit of common anode is the most popular circuit implementation due to the fact that offers power independency using an external power supply. The goal of the current application is to give two different solutions for using the SSD units (common anode or common cathode) that may be chosen by the engineer.

In the circuit of figure 6.8b, each segment activation is performed with a logic signal 0 (LOW), and the power supply is based on an external source. For supporting the above operation, the transistor BC557 (PNP type) is used instead of the BC547 (NPN type).

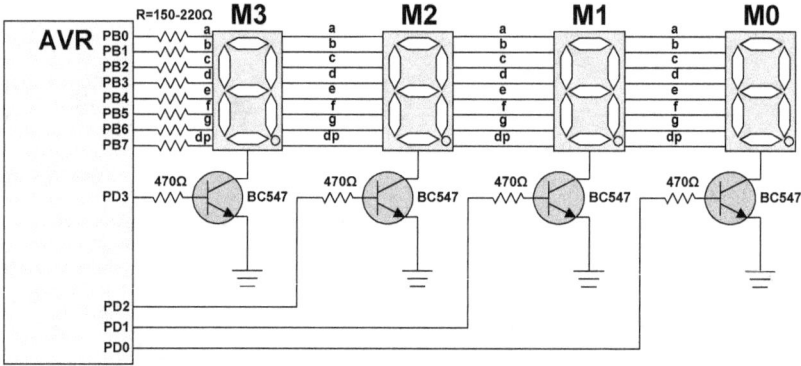

Figure 6.8a Circuit with four common cathode SSD units

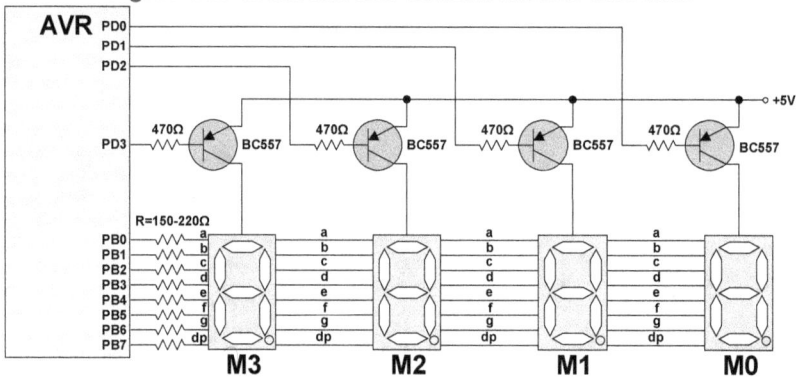

Figure 6.8b Circuit with four common anode SSD units

Now, more pins of port D are used (fig. 6.8a,b). Table 6.3 shows the pin values (PD3 to PD0) in order to display the digit '1' in all the SSD units (M3 to M0).

Table 6.3 SSD unit activation

PD3	PD2	PD1	PD0	Unit activation
1	0	0	0	M3
0	1	0	0	M2
0	0	1	0	M1
0	0	0	1	M0

As mentioned before, the display of every digit is performed in different stages as follows:
(a) display the digit, (b) deactivate the SSD unit and proceed to the next digit.
The operation of displaying and deactivating the SSD unit will be repeated many times and thus will be included in a macroinstruction. Initially, a microinstruction will be developed for activating the desired SSD unit based on a parameter that will be passed through the code. The name of the new macroinstruction is *setclear* and will be implemented by using the following code:

Code 6.4a

```
.MACRO setclear        ;Macroinstruction setclear

SBI PORTD,@0           ;Activate the pin @0 (parameter)
                       ;of port D (ground switch closed)
nop                    ;1 machine cycle (MC) delay
nop                    ;1 machine cycle (MC) delay
nop                    ;1 machine cycle (MC) delay

CBI PORTD,@0           ;Deactivate the pin @0 (parameter)
                       ;of port D (ground switch open)
.ENDMACRO              ;End of macroinstruction
```

Table 6.4 shows some examples of using the macroinstruction.

Table 6.4 SSD unit activation

PD3	PD2	PD1	PD0	Display unit (deactivated in the next stage)	Macroinstruction Unit
1	0	0	0	M3	setclear 3
0	1	0	0	M2	setclear 2
0	0	1	0	M1	setclear 1
0	0	0	1	M0	setclear 0

The time between the SSD units activation and deactivation still remains a question. This duration must be enough in order the digits to be readable by the human eye. According to literature, this duration must be less than 50mSec (50mSec equals to the frequency of 1/50mSec=20Hz).

The complete code follows.

Code 6.4b

```
;Place here the proper INC file for your microcontroller model
;(if needed), e.g. ATmega32/ATmega32A => m32def.inc/m32Adef.inc,
;ATmega328 => m328def.inc

.INCLUDE "include/m32Adef.inc"

.MACRO setclear        ;Macroinstruction setclear
```

```
        SBI  PORTD,@0              ;Activate the pin @0 (parameter)
                                   ;of port D (ground switch closed)

        nop                        ;1 machine cycle (MC) delay
        nop                        ;1 machine cycle (MC) delay
        nop                        ;1 machine cycle (MC) delay
        CBI  PORTD,@0              ;Deactivate the pin @0 (parameter)
                                   ;of port D (ground switch open)
        .ENDMACRO                  ;End of macroinstruction

;Initialization (setup outputs)
        LDI  R18,0xFF              ;Load the value for outputs
        OUT  DDRB,R18              ;Set all the pins of port B as outputs

        SBI  DDRD,0                ;Set PD0 as output
        SBI  DDRD,1                ;Set PD1 as output
        SBI  DDRD,2                ;Set PD2 as output
        SBI  DDRD,3                ;Set PD3 as output

start:                             ;Main program

;***************************
; Display the digit '0'
;***************************
        LDI  R24,0x3F              ;Load in R24 the segment bits
                                   ;for displaying the digit '0'
        OUT  PORTB,R24             ;Output to port B
        setclear 3                 ;Momentary activation of the M3 unit

;***************************
; Display the digit '1'
;***************************
        LDI  R24,0x06              ;Load in R24 the segment bits
                                   ;for displaying the digit '1'
        OUT  PORTB,R24             ;Output to port B
                                   ;Momentary activation of the M2 unit

;***************************
; Display the digit '2'
;***************************
        LDI  R24,0x5B              ;Load in R24 the segment bits
                                   ;for displaying the digit '2'
        OUT  PORTB,R24             ;Output to port B
                                   ;Momentary activation of the M1 unit

;***************************
; Display the digit '3'
;***************************
        LDI  R24,0x4F              ;Load in R24 the segment bits
                                   ;for displaying the digit '3'
        OUT  PORTB,R24             ;Output to port B
                                   ;Momentary activation of the M0 unit

        RJMP start                 ;Return to start
```

Figure 6.8c shows the four digits on the SSD units.

Figure 6.8c Four digits on the SSD units

Application 6.5 – Automated three digit number display

In previous application, the operation of displaying a digit in a specific SSD unit has been presented. The above operation is not suitable if the whole content of a register has to be displayed. If for example, the content of a register is 139, then the digits '1', '3' and '9' have to be displayed in different SSD units. Moreover, the above digits constitute a whole arithmetic value inside the register. Thus, the digits '1', '3' and 9 have to be separated from the whole value and then to be displayed on the SSD units. The digit separation can be achieved by successive divisions with the corresponding base of the arithmetic system which is the number 10. Every quotient (if is not zero), will be divided with 10. Every remainder represents a different digit of the initial number. Figure 6.9 shows the successive divisions for producing the separated digits of the number 139.

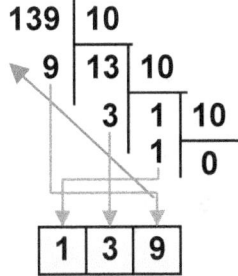

Figure 6.9 Digits separation with successive divisions

The resulting digits (division remainders) are produced in invert order as compared to the initial number. This has to be taken in account for displaying the number in the right form.

On the other hand, the AVR microcontroller does not have an instruction for division. Division can be performed only by 2 using the shift operation. For dividing with a different divisor (number 10 in the current application), the corresponding code has to be developed. It is known that the only core arithmetic operation of the microcontroller is the addition. Even the subtraction is performed by addition (using the two's complement).

Implementing the division algorithm

The good news is that the desired division can be performed by successive subtractions. The above procedure can be simplified because the AVR microcontroller has a subtraction instruction.

In this section, the division operation through successive subtractions will be presented and analyzed. Initially, the divisor is subtracted from the dividend. The operation is continued by subtracting the divisor from each new result. When the result is zero or negative the operation is terminated.

As an example, it is assumed that the division 139/10 will be calculated (fig. 6.10). The subtractions are stopped when the result is zero or negative. The negative result is not used. In this example (initial number 139), 13 subtractions have been performed and the last result was 9. Thus, the integer division 139/10 gives a quotient 13 (13 subtractions) and a remainder 9 (this is the first digit which is separated from the initial number). If the initial number (divisor) can be exactly divided by divisor, then the calculations are stopped when the result is zero.

139-10=129
129-10=119
119-10=109
109-10=99
99-10=89
89-10=79
79-10=69
69-10=59
59-10=49
49-10=39
39-10=29
29-10=19
19-10=9
9-10=-1 (<0)

Figure 6.10 Successive subtractions

As a second example, the calculation of the division 6/2 is analyzed.

Step 1 6-2=4
Step 2 4-2=2
Step 3 2-2=0 (the result is zero and the calculation is terminated)

Based on the above calculations, the quotient is 3 (three subtractions are performed) and the remainder is 0.

If the result is negative, then the remainder is the last non negative result. If the result is zero, then the remainder is equal to this result. A negative result is always not acceptable.

Figure 6.11 shows the algorithm for implementing the division (through subtraction) for calculating the quotient and the remainder.

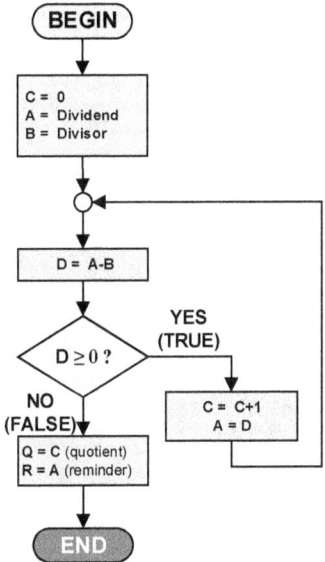

Figure 6.11 Basic division algorithm

The next step is to verify the correctness of the above algorithm using the calculation 139/10.

Integer division 139/10	
Step 1 C=0, A=139, B=10 D=139-10=129 D>=0, C=1, A=129 Step 2 D=129-10=119 D>=0, C=2, A=119 Step 3 D=119-10=109 D>=0, C=3, A=109 Step 4 D=109-10=99 D>=0, C=4, A=99 Step 5 D=99-10=89 D>=0, C=5, A=89 Step 6 D=89-10=79	Step 8 D=69-10=59 D>=0, C=8, A=59 Step 9 D=59-10=49 D>=0, C=9, A=49 Step 10 D=49-10=39 D>=0, C=10, A=39 Step 11 D=39-10=29 D>=0, C=11, A=29 Step 12 D=29-10=19 D>=0, C=12, A=19 Step 13 D=19-10=9 D>=0, C=13, A=9

| D>=0, C=6, A=79
Step 7
D=79-10=69
D>=0, C=7, A=69 | Step 14
D=9-10=-1
D<0, Q=C=13, R=A=9 |

The above algorithm performs only one division. This means that for implementing more divisions, the algorithm has to be 'executed' more times.
Figure 6.12 shows the division algorithm (as subroutine) which is adapted to the final implementation (for performing multiple divisions).

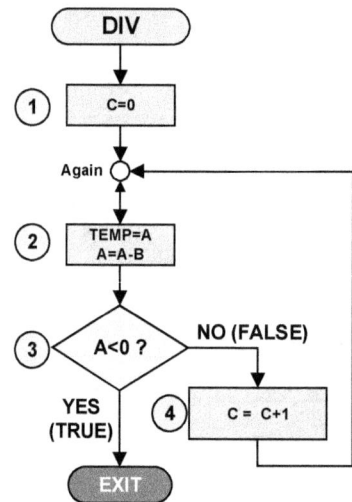

Figure 6.12 Final algorithm for implementing division

The following code implements the above division algorithm.

Code 6.5a

```
DIV:
(1)   LDI  C,0              ;Subtractions counter
      Again:
(2)        MOV  TEMP,A      ;Temporarily storage of dividend
           SUB  A,B         ;Divisor subtraction
(3)        BRCS exit        ;If the result is negative,
                            ;then the calculation is terminated
(4)        INC  C           ;Increment subtractions counter
           RJMP Again       ;Return for the next subtraction
      exit:
                            ;After the subroutine exit,
                            ;C=quotient
                            ;TEMP=reminder
RET
```

The operation of the above code is analyzed as follows:

```
        LDI C,0
```
The variable C (through a .DEF directive inside code) counts the number of subtractions (with positive or zero result) and is equivalent to the quotient.

```
        Again:
```
Reference point (address) for the next subtraction

```
        MOV TEMP,A   ; Temporarily storage of dividend
```
The content of A is stored temporarily in the variable TEMP. This is performed because the content of A is overwritten by the new result of abstraction. When a negative result is occurred, the last content of A is the last remainder. It must be noticed that the result from each subtraction does not represent the final remainder of the division. The final remainder is represented only from the last accepted result. When the current result is negative (not acceptable result), the last accepted result is stored in the TEMP variable.

```
        SUB A,B      ; Divisor subtraction
```
The above instruction calculates each new result from the abstraction.

```
        BRCS exit    ; If the result is negative,
                     ; then the calculation is terminated
```
In a case of negative result, the whole procedure is terminated, the C variable is not updated and the execution flow does not return to the Again point.

```
        INC C        ; Increment subtractions counter
```
If the previous result is not negative, then the subtractions counter is increased.

```
        RJMP Again
```
Return for the next subtraction

```
        exit:
```
label for avoiding the previous two instructions and terminating the operation. After the exit point, the quotient is stored to the variable C and the remainder to the variable TEMP.

For reusing the above code, a subroutine with the name DIV is created.
The 'variables' A,B,C and TEMP represent registers (using the directive .DEF).

Three digit number separation
The above subroutine DIV will be called within a loop for processing any number of digits. Figure 6.13 shows the corresponding algorithm.

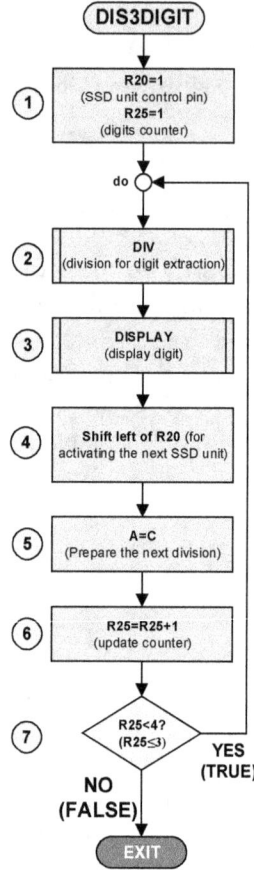

Figure 6.13 Using the division inside a loop

In the following code, the DIV subroutine is called repetitively in order to separate a three digit number to the corresponding digits. These digits are displayed on the SSD units.

Code 6.5b

```
DIS3DIGIT:
(1) LDI R20,1           ;Pin number for activating
                        ;a specific SSD unit
    LDI R25,1           ;Initialize counter (number of digits)
    do:                 ;Return point (Loop)
(2) RCALL DIV           ;Integer division
(3) RCALL DISPLAY       ;Display digit
(4) LSL R20             ;Left shift for activating the
                        ;next SSD unit on the left
(5) MOV A,C             ;The new number to be divided
(6) INC R25             ;Increase the digits counter
(7) CPI R25,4
    BRLT do             ;While all the digits have not be displayed
RET
```

The above code is analyzed as follows:

```
    LDI R20,1           ;Pin number for activating
                        ;a specific SSD unit
```
The signal level of three pins of port D, defines which of the three SSD units will be activated. The activation of the three SSD units is mutual excluded and thus, the only possible signal values of port D are:

00000001 (1 for the pin PD0) for activation of the SSD unit M0 (figure 6.8a)
00000010 (1 for the pin PD1) for activation of the SSD unit M1 (figure 6.8a)
00000100 (1 for the pin PD2) for activation of the SSD unit M2 (figure 6.8a)

```
    LDI R25,1           ;Initialize counter (number of digits)
```
The counter R25 is used for counting the digits that will be displayed. For three digits, the value range is 1 to 3.

```
    do:                 ;Return point (Loop)
```
Reference point for displaying the next digit.

```
    RCALL DIV           ;Integer division
```
Separation of the first digit from the right.

```
    RCALL DISPLAY       ;Display digit
```
Call the subroutine for displaying the digit.

```
    LSL R20             ;Left shift for activating the
                        ;next SSD unit on the left
```
After displaying the digit, a shift of R20 content is performed in order to activate the next SSD unit. For example, if the previous content of R20 is 00000001, then with a left shift, the content becomes 00000010.

```
    MOV A,C             ;The new number to be divided
```
The next division will be performed by using the previous quotient as dividend.

```
    INC R25             ;Increase the digits counter
```
The content of the register R25 is increased (one more digit has been displayed)

```
    CPI R25,4
```
Compare the counter content (R25) with the value 4 (see in the next instruction the branch type)

```
    BRLT do             ;While all the digits have not be displayed
```
Return to the do point, if R25≤3 (R25<4)

Displaying digits on the SSD units

As mentioned before, an array (sequence) of hexadecimal bytes has to be created in order to lighting up the corresponding LEDs on the SSD unit for representing a specific digit. This array starts from the symbolic address `digits`. The digit that will be displayed, is the division remainder which is stored in TEMP variable after the termination of the operation. Thus, the TEMP content is used as distance pointer from the symbolic address digits in order to send to port B the right bits. Figure 6.14 shows the flow chart diagram for the above operation.

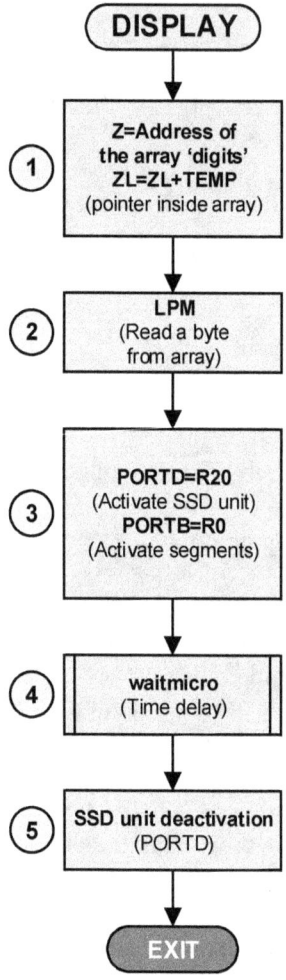

Figure 6.14 Display algorithm

The following code represents the operation for displaying a digit to a specific SSD unit by applying the right bits on the pins of ports B and D.

Code 6.5c

```
DISPLAY:
(1) LDI  ZL,LOW(digits*2)    ;Initialize register Z
    LDI  ZH,HIGH(digits*2)   ;in the digits array
    ADD  ZL,TEMP             ;Update the distance pointer
                             ;for reading from the digits array
```

```
(2) LPM                     ;Read one byte from the address
                            ;which is represented by the register Z
                            ;Store the byte in R0 (Low Byte)

(3) OUT PORTD,R20           ;Activate the SSD unit
    OUT PORTB,R0            ;Activate segments
(4) RCALL waitmicro         ;A short time delay
                            ;(hundreds of µSec for a clock
                            ;frequency of 1MHz)

(5) LDI R26,zero            ;Initialize R26
    OUT PORTD,R26           ;Deactivate segments
RET
```

The operation of the above code is analyzed as follows:

```
LDI   ZL,LOW(digits*2)    ; Initialize register Z
LDI   ZH,HIGH(digits*2)   ; in the digits array
```
Load in register Z (16bit) the starting address of digits array

```
ADD ZL,TEMP    ; Update the distance pointer
               ; for reading from the digits array
```
In Low byte of the address, the remainder value from the integer division is added. If for example, TEMP=2, then the address digits+2 is calculated, in order to display the digit '2'.

```
LPM                  ; Read one byte from the address
                     ; which is represented by the register Z
                     ; Store the byte in R0 (Low Byte)
```
Read a byte from the program memory (address Z). This byte is stored in the register R0 (the bit values activate the corresponding segments of the SSD unit).

```
OUT PORTD,R20     ; Activate the SSD unit
```
Activation of a specific SSD unit based on the signal values of port D.

```
OUT PORTB,R0      ; Activate segments
```
Activate the corresponding pins of port B for lighting up the right segments of the SSD unit.

```
RCALL waitmicro     ; A short time delay
                    ; (hundreds of µSec for a clock
                    ; frequency of 1MHz)
```
Time delay before the segment deactivation

```
LDI R26,zero        ;Initialize R26
```
Clear the register content (a clear instruction has not been used for initializing the register content with any desired value)

```
OUT PORTD,R26     ; Deactivate segments
```
Deactivate SSD unit through the port D.

Other supplementary subroutines

In addition from the above subroutines, a time delay (active time duration for displaying a digit on a SSD unit) and a initialization procedure have been developed. The flow chart diagrams of figures 6.15 and 6.16 show the above subroutines (waitmicro and INIT).

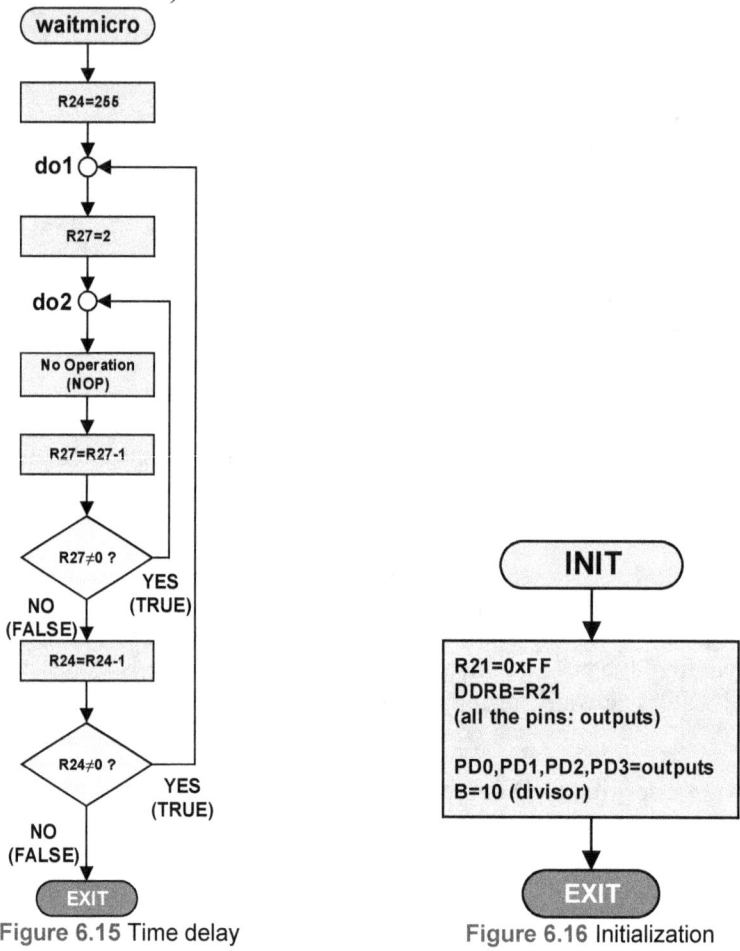

Figure 6.15 Time delay   Figure 6.16 Initialization

Figure 6.17 shows the completed flow chart diagram of the whole operation.

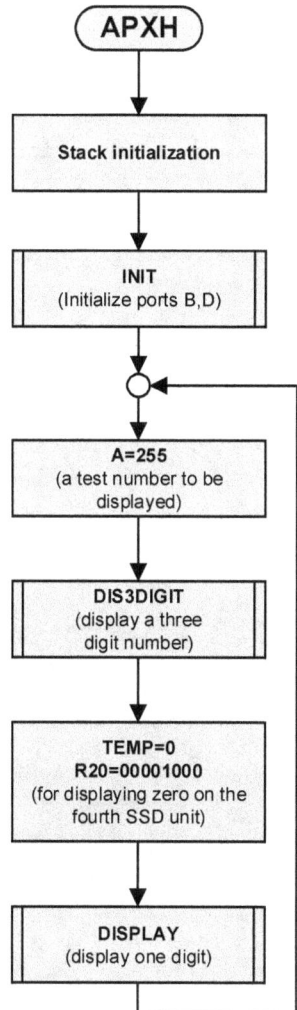

Figure 6.17 The main application algorithm

The completed code is as follows:
Code 6.6

```
;Place here the proper INC file for your microcontroller model
;(if needed), e.g. ATmega32/ATmega32A => m32def.inc/m32Adef.inc,
;ATmega328 => m328def.inc

.INCLUDE "include/m32Adef.inc"

;**************************
;Stack initialization
;**************************
LDI R16,HIGH(RAMEND)
OUT SPH,R16
LDI R16,LOW(RAMEND)
OUT SPL,R16
```

```
;***************************
;Supplementary declarations
;***************************
    .DEF C=R16          ;"replace" registers with symbolic names
    .DEF A=R17          ;for making the divisions easier
    .DEF B=R18
    .DEF TEMP=R19

    .EQU zero=0x00      ;A constant for the 0x00 value

    RCALL INIT          ;Call the subroutine for setting ports
                        ;and divisor

;***************************
;Main program
;***************************
restart:                ;Restart point for refreshing display

    LDI A,255           ;255=the test number to be displayed
                        ;on the three SSD units

    RCALL DIS3DIGIT     ;Call the subroutine for displaying
                        ;the three digits number

    LDI TEMP,0          ;Digit to be displayed
    LDI R20,0b00001000  ;Activate the first (from the left) SSD unit

    RCALL DISPLAY       ;Call the subroutine for displaying one digit

    RJMP restart        ;Return to the restart point

;*******************************
;Integer division subroutine
;*******************************
  DIV:
    LDI C,0                     ;Subtraction counter
        Again:
            MOV TEMP,A          ;Temporary storage of the dividend
            SUB A,B             ;Subtract the divisor
            BRCS exit           ;If the result is negative, then the
                                ;operation is terminated

            INC C               ;Increase the subtractions counter
            RJMP Again          ;Return for the next subtraction
        exit:
                                ;After the termination of the operation
                                ;C=quotient
                                ;TEMP=remainder
RET

;*******************************
;Three digits display
;*******************************
DIS3DIGIT:
LDI R20,1                       ;Pin number for the SSD unit
                                ;that will be activated
```

```
    LDI R25,1              ;counter (number of digits)
  do:                      ;Loop restart point
    RCALL DIV              ;Integer division (first digit on the
                           ;right)
    RCALL DISPLAY          ;Display digit
    LSL R20                ;Left shift for activating the next SSD
                           ;unit (on the left)
    MOV A,C                ;The new number that will be divided
                           ;is the produced quotient
    INC R25                ;Increase counter (number of digits)
    CPI R25,4              ;If all the digits have not been
                           ;displayed, then return to loop
                           ;starting point
    BRLT do
RET

;******************************
;Digit display
;******************************
DISPLAY:

  LDI ZL,LOW(digits*2)     ;Initialize register Z
  LDI ZH,HIGH(digits*2)    ;in the digits array
  ADD ZL,TEMP              ;Update the distance pointer
                           ;for reading from the digits array
  LPM                      ;Read one byte from the address
                           ;which is represented by the register Z
                           ;Store the byte in R0 (Low Byte)

  OUT PORTD,R20            ;Activate the SSD unit
  OUT PORTB,R0             ;Activate segments
  RCALL waitmicro          ;A short time delay
                           ;(hundreds of µSec for a clock
                           ;frequency of 1MHz)
  LDI R26,zero             ;Initialize R26
  OUT PORTD,R26            ;Deactivate segments
  RET

;******************************
;Port initialization
;******************************
INIT:
LDI R21,0xFF
OUT DDRB,R21

SBI DDRD,0
SBI DDRD,1
SBI DDRD,2
SBI DDRD,3
LDI B,10
RET

;******************************
;Time delay in µSec
;******************************
waitmicro:
```

```
        LDI R24,255
    do1:
            LDI R27,2
        do2:
                nop
                nop
                DEC R27
            BRNE do2
            DEC R24
        BRNE do1
RET

digits:  .DB 0x3F,0x06,0x5B,0x4F,0x66,0x6D,0x7D,0x07,0x7F,0x6F
```

Figure 6.18 shows the number 255 on the SSD unit.

Figure 6.18 The number 255 on the SSD unit

6.2 LCD Screen

The display operation is critical for the application developer as well as for the end user. The information display is also used to verify the application operation by the developer. The LCD screens have some significant advantages as compared to the SSD units. Some of the advantages, are:

(a) a full text can be displayed
(b) the LCD screen is controlled via an embedded microcontroller. This feature offers greater programming capabilities for the LCD screen. Moreover, the LCD screen can be directly connected to the microcontroller without any additional circuits.
(c) the purchase cost is low
(d) can be used with any microcontroller

On the other hand, the SSD units are high bright units (have LEDs inside) and are more suitable for viewing them in limited light conditions or from distance.
Thus, based on the application requirements, the most suitable solution will be chosen by the engineer. Figure 6.19 shows the physical form of an LCD screen 16x2 (16 columns x 2 rows).

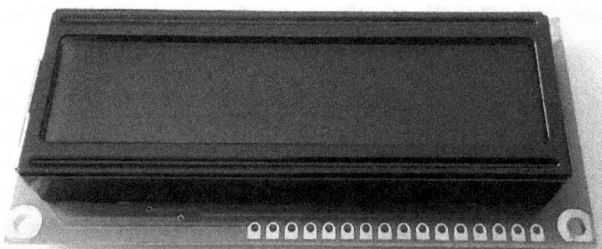

Figure 6.19 The LCD screen 16x2

Figure 6.20 shows the LCD pins (16x2).

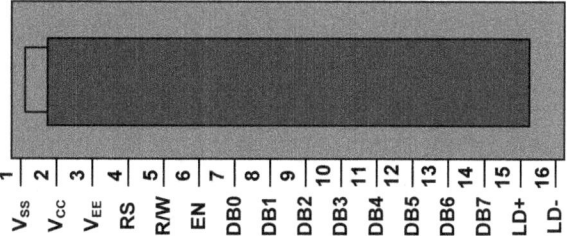

Figure 6.20 LCD 16x2 pins

Table 6.5 shows the LCD pin usage.

Table 6.5 Pin usage

Pin number	Symbolic name	Usage
1	V_{SS}	Ground
2	V_{CC}	+5V (power supply)
3	V_{EE}	Contrast control
4	RS	Register selection (0=instruction register, 1= data register)
5	R/W	Read/Write selection (0=write, 1=read)
6	EN	Operation activation
7-14	DB0-DB7	Data lines (bus)
15	LD+	+5V for backlight*
16	LD-	Ground for backlight*

*Some LCD types do not support backlight and thus the available pins are 14 and not 16.

The LCD microcontroller has two 8bit registers:

IR – Instruction Register
DR – Data Register

IR is used for controlling the LCD operation (e.g. clear screen, text shift). DR is used for the data regarding the screen memory. The LCD system includes two types of memory.

DDRAM – Display Data RAM and CGRAM – Character Generator RAM

The DDRAM is related with the visible characters on the screen. In other words, if a character is stored within this memory, then is also visible on the screen. As mentioned before, 16 characters (16 columns) can be displayed in two rows. Figure 6.21 shows the corresponding addresses which represent the display locations on the screen.

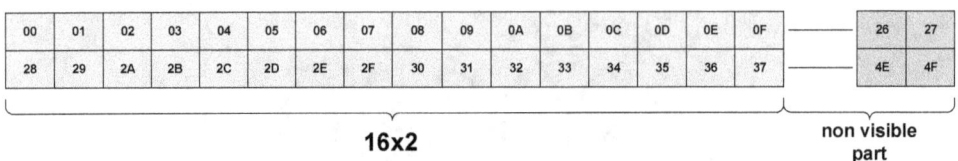

Figure 6.21 LCD screen DDRAM

As shown in figure 6.21, the visible characters correspond to the addresses 00 to 0F (hexadecimal) for the first row and 28 to 37 (hexadecimal) for the second row. The DDRAM consists of 80 locations of 8bits. That means that when more than 16 characters are sent to the LCD screen, the characters after the 16th are not displayed. On the other hand, the CGRAM stores ASCII characters for direct displaying of messages, while new symbols can be created for replacing the existing.

The code-number of some basic LCD instructions is shown in the table 6.6.

Table 6.6 Instructions (or commands) to the LCD

Code (hexadecimal)	Operation
1	Clear screen
4	Decrement cursor (shift cursor to left)
6	Increment cursor (shift cursor to right)
5	Shift right (screen content)
7	Shift left (screen content)
8	Clear screen, cursor off
A	Clear screen, cursor on
C	Activate screen, cursor off
E	Activate screen, cursor blink
10	Shift cursor position to left
14	Shift cursor to right
80	Move the cursor at the beginning of the first row
C0	Move the cursor at the beginning of the second row

The circuit for controlling the LCD screen is shown in figure 6.22. The screen contrast is adjusted through a 10KΩ potentiometer which is connected with power supply (left and right pin) and with the V_{EE} using the middle pin. In current circuit, 8 pins of the microcontroller are used for data (symbols to be displayed), while the control signals RS, R/W and EN are connected to a different port.

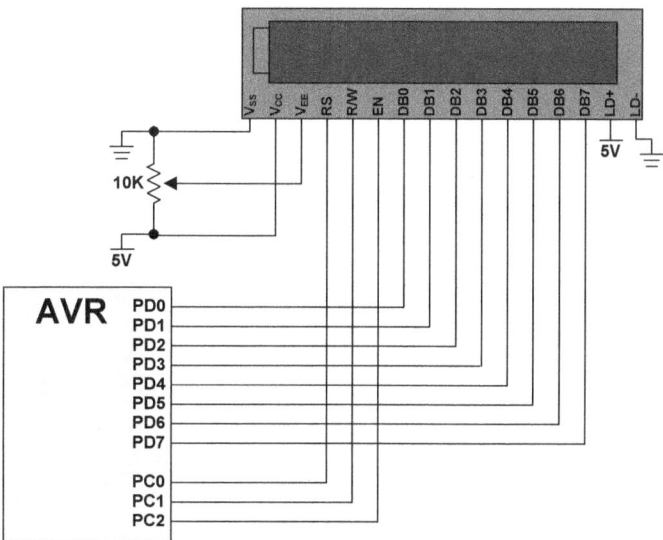

Figure 6.22 Microcontroller-LCD connection

Application 6.6 – Using the LCD screen (16x2)

In this application, the circuit of figure 6.22 will be implemented. Figure 6.23 shows the text messages on the LCD screen (in two rows). This application is developed as an example in order to present the basic LCD screen manipulation.

Figure 6.23 Text messages on the LCD

For the LCD normal operation, the proper signals from the microcontroller have to be sent. Figure 6.24 shows the timing diagram of signal sequences in order to send instructions and symbols to the LCD screen. The signal transitions must fulfil some minimum requirements in order to produce the correct LCD screen response. Otherwise, the text symbols will not be displayed correctly. In practical level, time delays must be introduced before any signal change in the microcontroller outputs. This is necessary, because the LCD screen embedded microcontroller needs time spaces in order to complete some operations. For example a time space of 1.6mSecs is needed for clearing the screen. Thus, any new instruction after the LCD clearing operation has to be executed after a time delay of 1.6mSecs. Another critical time interval, is the switching process between LCD instruction and LCD data (symbols to be displayed). In this case, a time delay is also needed. The timing requirements for the LCD operation are presented in the corresponding Datasheet of the manufacturer. Figure 6.24 shows the sequence and the corresponding signal transitions in order to display the message "LCD 16x2". This diagram presents also the initialization and the activation phases of the LCD screen. It must be noticed that the horizontal axis does not represent a real time period, because the diagram is only focused on the signal forms and transitions.

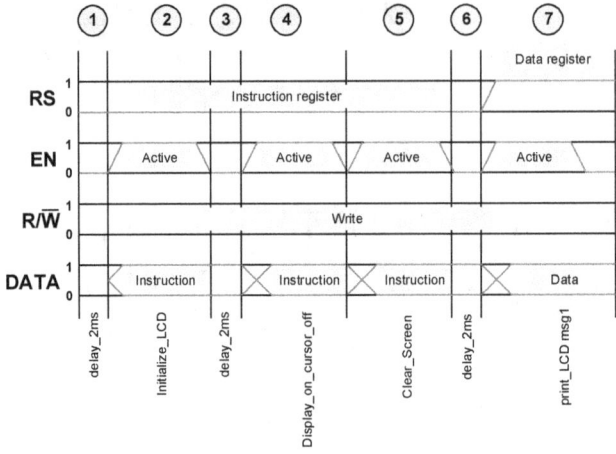

Figure 6.24 Signals for the LCD operation

Based on figure 6.24, the required signals belong to the following phases:
(1) Wait (2mSecs) for LCD circuit initialization
(2) RS=0 (Instruction register selected), EN=1 (activation), R/W=0 (write), DATA=Instruction code for LCD initialization
(3) Wait (2mSec) before the normal instruction sequence
(4) Instruction for LCD activation and cursor off
(5) Instruction for clearing the LCD
(6) Wait (2mSecs) before switching for sending data (symbols to be displayed)
(7) Display the first message

Based on the above, it is obvious that two types of write operations exist:

(a) Write instruction
(b) Write data (symbols to be displayed)

Figures 6.25 and 6.26 show the two basic operations that will be implemented for sending instructions and data to the LCD.

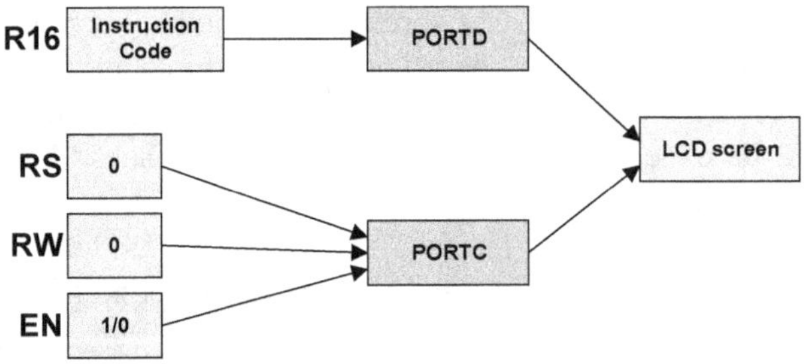

Figure 6.25 Writing instructions

Figures 6.25 and 6.26, show how the register R16 is used for loading the hexadecimal number which will be sent to output. The only difference between these figures (fig. 6.25, 6.26) is the RS value for selecting the target register.

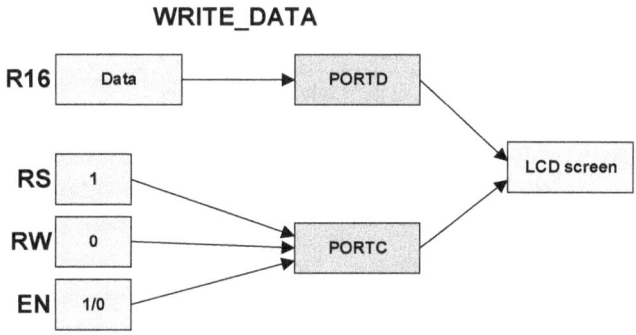

Figure 6.26 Writing Data

The required code has to generate the proper signals in order to display the messages on the LCD screen. Figure 6.27 shows the signal forms which are generated from the corresponding code for sending the instruction (SEND_COMMAND).
In the same figure (fig. 6.27) the signals until the display of the first text row are presented.

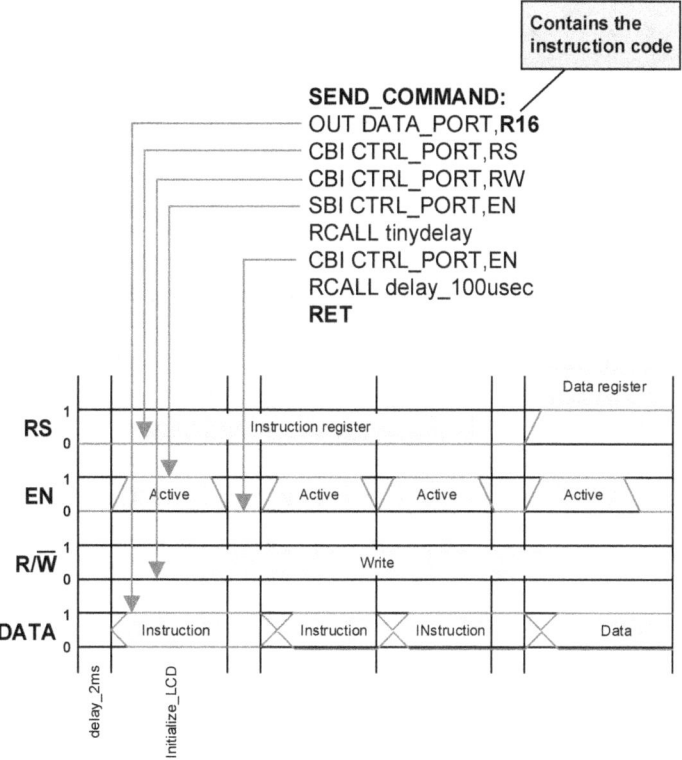

Σχήμα 6.27 Generated signals from the code SEND_COMMAND

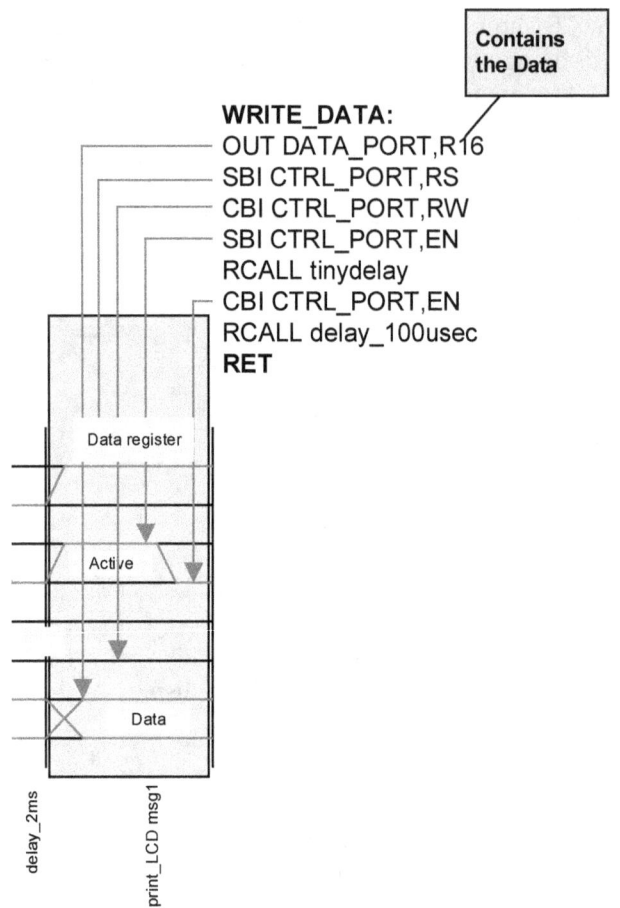

Figure 6.28 Generated signals from the code WRITE_DATA

Figure 6.28 represents the writing operation. It must be noticed that the arrows in the above figures show only the signals and instructions and not the exactly point of time where they generated.

The operation for displaying both text messages is always the same. The only difference is the corresponding memory area which represents each LCD row.

Thus, the text messages will be stored inside program memory (inside code) and the whole display operation will be implemented within a macroinstruction.
Figures 6.29 and 6.30 shows the activation of the macroinstruction print_LCD for displaying the messages msg1 and msg2.

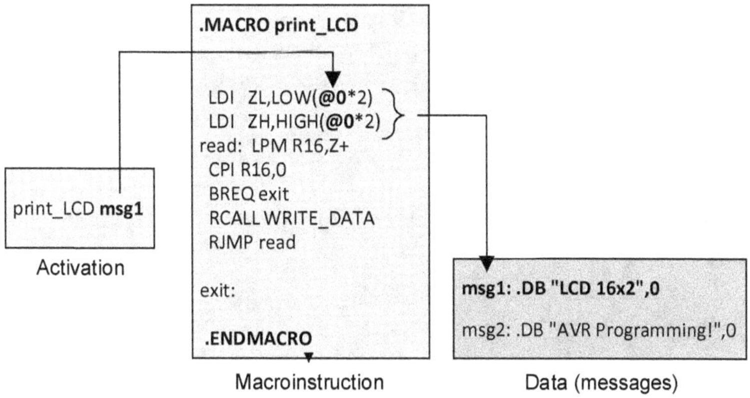

Figure 6.29 Displaying the message "LCD 16x2"

The symbolic names `msg1` or `msg2` are passed as parameters and the corresponding code reads the message bytes.

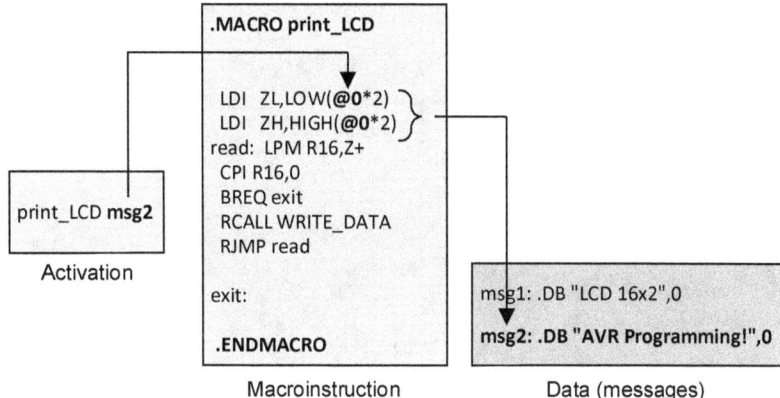

Figure 6.30 Displaying the message "AVR Programming!"

Figure 6.31 shows the application flow chart diagram.

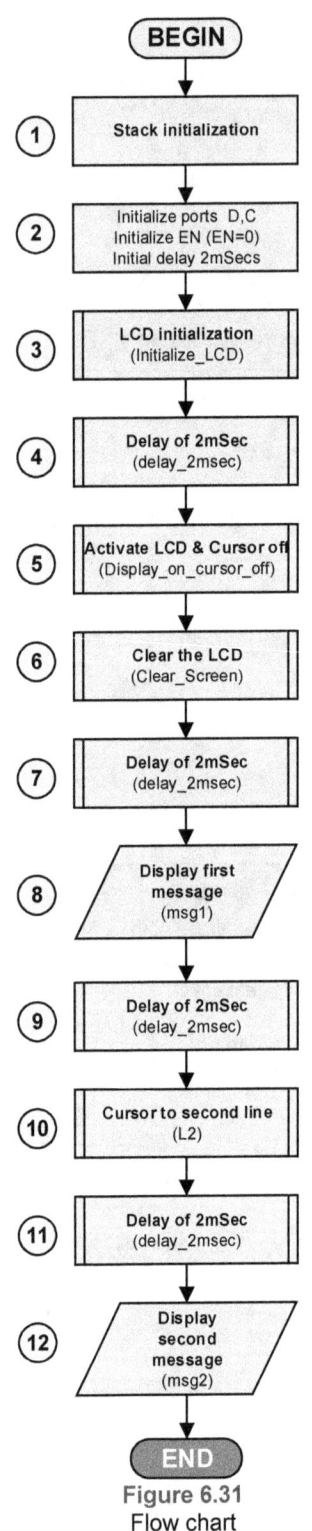

Figure 6.31
Flow chart

Code development
Before the main section development, some symbolic names will be declared for simplifying the code.

```
.EQU DATA_DDR= DDRD
.EQU DATA_PORT = PORTD
.EQU DATA_PIN = PIND
```

Port D is used for transmitting data to the pins DB0 to DB7.

```
.EQU CTRL_DDR = DDRC
.EQU CTRL_PORT = PORTC
```

Port C is used for the control signals RS, RW and EN.

```
.EQU RS = 0
.EQU RW = 1
.EQU EN = 2
```

Port C pin number.

(1)
```
LDI R16,HIGH(RAMEND)
OUT SPH,R16
LDI R16,LOW(RAMEND)
OUT SPL,R16
```

The stack initialization is performed for preparing the operations of call and return instructions from the subroutines.

(2)
```
LDI R16,0xFF
OUT DATA_DDR, R16
OUT CTRL_DDR, R16
CBI CTRL_PORT,EN
RCALL delay_2msec
```

Set ports D and C as outputs, initialize the signal EN and set the initial delay.

(3)
```
RCALL Initialize_LCD
```
LCD initialization

(4)
```
RCALL delay_2msec
```
Time delay of 2mSecs

(5)
```
RCALL Display_on_cursor_off
```
Activate LCD screen and cursor off

(6)
```
RCALL Clear_Screen
```
Clear the LCD screen

(7)
```
RCALL delay_2msec
```
Time delay of 2mSecs

(8)
```
print_LCD msg1
```
Display the first message

(9)
```
RCALL delay_2msec
```
Time delay of 2mSecs

(10)
```
RCALL L2
```
Move the cursor at the beginning of the second row for displaying later the second message.

(11)
```
RCALL delay_2msec
```
Time delay of 2mSecs

(12)
```
print_LCD msg2
```
Display the second message

```
END: RJMP END
```
Program termination (the execution flow is trapped here)

The following code represents the subroutines `SEND_COMMAND` and `WRITE_DATA`.

```
tinydelay:

NOP
```
A short delay (one machine cycle)

```
RET
```
Return from subroutine

```
delay_100usec:
PUSH R17
LDI R17,60
again:
      RCALL tinydelay
      DEC R17
BRNE again
POP R17
RET
```

A time delay of approximately 100μSecs (using the subroutine `tinydelay`). It must be noticed that the current value of R17 register is stored in the stack. This is done because the loops of the subroutines `delay_100usec` and `delay_2msec` use the same register as counter. Thus, the current value is stored initially in the stack and after the next loop termination, is recalled.

```
delay_2msec:
PUSH R17
LDI R17,20
again2:
        RCALL delay_100usec
        DEC R17
BRNE again2
POP R17
RET
```

The above loop uses the subroutine `delay_100usec` for achieving a time delay of 2mSecs.

In the following code, the only difference is the instruction code which represents a different operation.

```
Display_on_cursor_off:
  LDI R16,0x0C
  RCALL SEND_COMMAND
RET

Initialize_LCD:
  LDI R16,0x38
  RCALL SEND_COMMAND
RET

Clear_Screen:
  LDI R16,0x01
  RCALL SEND_COMMAND
RET

L2:
  LDI R16,0xC0
  RCALL SEND_COMMAND
RET

msg1: .DB "LCD 16x2",0
msg2: .DB "AVR Programming!",0
```
Message declaration inside the program memory

As a final step, a macroinstruction has been developed for displaying the messages:

```
.MACRO print_LCD

   LDI    ZL,LOW(@0*2)
   LDI    ZH,HIGH(@0*2)
read:  LPM R16,Z+
   CPI R16,0
```

```
    BREQ exit
    RCALL WRITE_DATA
    RJMP read

exit:

.ENDMACRO
```

The instruction `LPM R16, Z+` loads in the register R16 the byte (ASCII code) that will be displayed on the LCD (successive bytes represent the message) and then, increases the content of the Z register which is used as a distance pointer inside the byte sequence of the messages `msg1` or `msg2`. The content of Z is increased in order to display the next character (byte) of the message. The current character (of the message) is compared with zero (`CPI R16, 0`) for detecting the end of message. If the current character is zero (the BREQ condition is true), then the display character operation is completed and the execution flow is driven outside the loop (label `exit`).

The complete application code is as follows:
Code 6.7

```
;Place here the proper INC file for your microcontroller model
;(if needed), e.g. ATmega32/ATmega32A => m32def.inc/m32Adef.inc,
;ATmega328 => m328def.inc

.INCLUDE "include/m32Adef.inc"

;*************************
;Stack initialization
;*************************
LDI R16,HIGH(RAMEND)
OUT SPH,R16
LDI R16,LOW(RAMEND)
OUT SPL,R16

;********************************
;Symbolic names declaration
;********************************
  .EQU DATA_DDR= DDRD
  .EQU DATA_PORT = PORTD
  .EQU DATA_PIN = PIND

  .EQU CTRL_DDR = DDRC
  .EQU CTRL_PORT = PORTC

  .EQU RS = 0
  .EQU RW = 1
  .EQU EN = 2

;********************************
;Macroinstruction
;********************************
  .MACRO print_LCD
```

```
   LDI    ZL,LOW(@0*2)
   LDI    ZH,HIGH(@0*2)
read: LPM R16,Z+
  CPI R16,0
  BREQ exit
  RCALL WRITE_DATA
  RJMP read

exit:

 .ENDMACRO

;*******************************
;Ports settings
;*******************************

LDI R16,0xFF
OUT DATA_DDR, R16
OUT CTRL_DDR, R16

;*******************************
;LCD initialization
;*******************************
CBI CTRL_PORT,EN

;Wait for initialization
RCALL delay_2msec

;*******************************
;Main program
;*******************************
RCALL Initialize_LCD
RCALL delay_2msec
RCALL Display_on_cursor_off
RCALL Clear_Screen
RCALL delay_2msec

print_LCD msg1
RCALL delay_2msec
RCALL L2
RCALL delay_2msec

print_LCD msg2

END: RJMP END

;*******************************
;Subroutines
;*******************************
                                  ;Write instruction
SEND_COMMAND:
      OUT DATA_PORT,R16
      CBI CTRL_PORT,RS
      CBI CTRL_PORT,RW
      SBI CTRL_PORT,EN
      RCALL tinydelay
```

```
        CBI CTRL_PORT,EN
        RCALL delay_100usec
RET

                                        ;Write data
WRITE_DATA:
        OUT DATA_PORT,R16
        SBI CTRL_PORT,RS
        CBI CTRL_PORT,RW
        SBI CTRL_PORT,EN
        RCALL tinydelay
        CBI CTRL_PORT,EN
        RCALL delay_100usec
RET

                                        ;A short delay
tinydelay:
        NOP
RET

                                        ;A delay of 100μSecs
delay_100usec:
PUSH R17
LDI R17,60
again:
        RCALL tinydelay
        DEC R17
BRNE again
POP R17
RET
                                        ;A delay of 2mSecs
delay_2msec:
PUSH R17
LDI R17,20
again2:
        RCALL delay_100usec
        DEC R17
BRNE again2
POP R17
RET

                                        ;Activate LCD, cursor off
Display_on_cursor_off:
  LDI R16,0x0C
  RCALL SEND_COMMAND
RET
                                        ;Initialization
Initialize_LCD:
  LDI R16,0x38
  RCALL SEND_COMMAND
RET
                                        ;Clear the LCD
Clear_Screen:
  LDI R16,0x01
  RCALL SEND_COMMAND
RET
```

```
L2:
    LDI R16,0xC0          ;Cursor at the second row
    RCALL SEND_COMMAND
RET

                          ;Messages declarations
msg1:   .DB "LCD 16x2",0
msg2:   .DB "AVR Programming!",0
```

LABORATORY EXERCISE 4
Seven Segment Display (SSD) units

GOAL
The goal of this exercise is to use in practice the SSD units which are used in many applications.

Step 1
Initially, the operation of a SSD unit will be tested by connecting manually each segment with a power supply through a resistor. It is known that each segment is represented by a LED and thus the proper connection has to be implemented. A common cathode (CC) SSD unit will be used. That means that the common pin will be connected to ground. For activating a LED (segment) the corresponding pin has to be connected to +5V. Connect successively every SSD unit segment and test the corresponding operation. For all the connections, resistors (e.g.220Ω) have to be used.

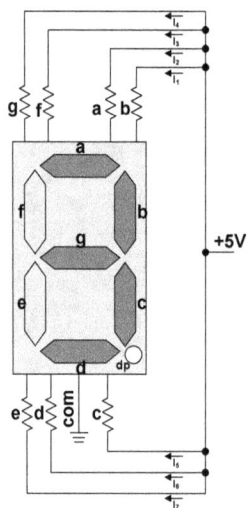

Step 2
(a) Calculate the current flow of each LED circuit
(b) Calculate the total current flow in order to display the digit '3' on the SSD unit
(c) Calculate the total current flow for activating all the SSD unit segments

Step 3
For manipulating multiple SSD units, the multiplexing method is used. Based on this method, a transistor is used as a switch. Every time a current is flown at the transistor base, the common cathode of the SSD unit is connected to ground and the LED is lit. Implement the following circuit.

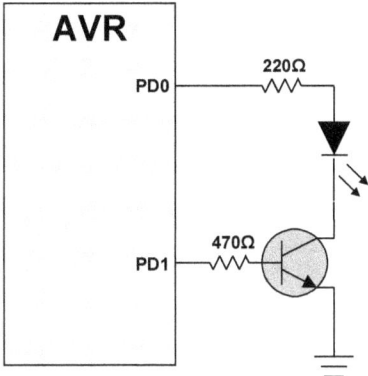

Develop a program to test all the possible signal combinations at the pins PD0 and PD1 and observe the results on the LED.

Step 4
Fill the following table based on the experimental tests of the previous circuit.

PD1	PD0	LED State
0	0	
0	1	
1	0	
1	1	

Step 5
Implement the following circuit.

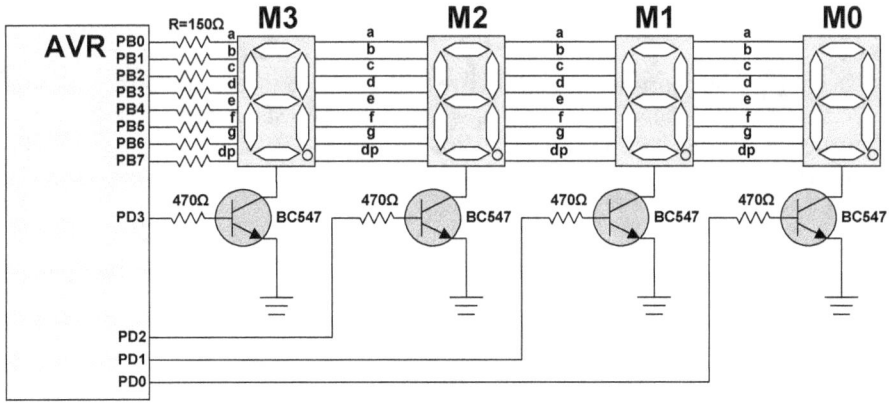

Step 6
Develop a program to display successively the digit '1' on the SSD units (M3 to M0) with a delay of 1 Sec.

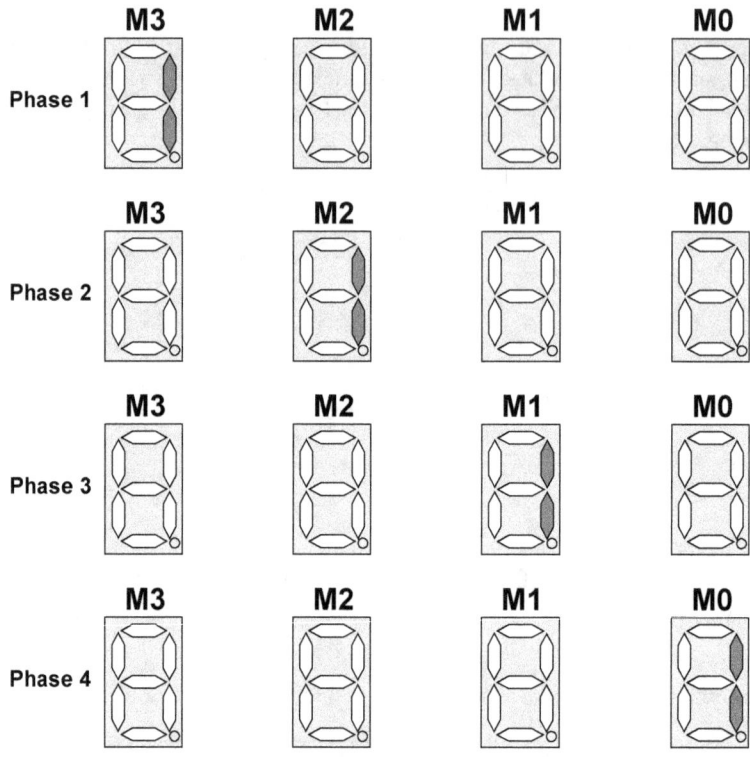

Step 7
Based on the previous phases, fill the following table:

Phase	PB0	PB1	PB2	PB3	PB4	PB5	PB6	PB7	PD0	PD1	PD2	PD3
1												
2												
3												
4												

Step 8
The following circuit constitutes an alternative version of the first circuit. In this version, SSD units of common anode are used. Thus the common SSD unit pin is connected to +5V, while the segment activation is achieved by using a 0V (logical zero) signal. Using the above circuit implementation, the needed power is supplied by an external source. The new circuit is as follows:

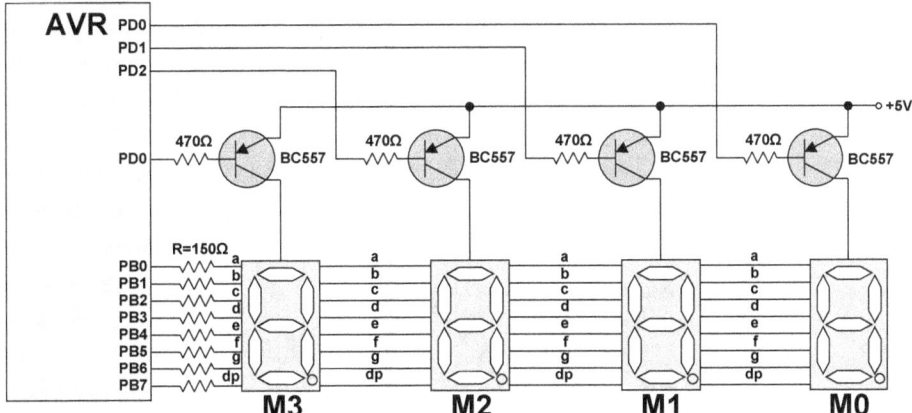

In the above circuit, a BC557 transistor is used (PNP type). Implement the above circuit and make comments based on the corresponding practical operation as compared to the common cathode version.

LABORATORY EXERCISE 5
Using an LCD 16x2 screen (1602)

GOAL
The familiarization with the classic LCD (16 columns and 2 rows), by exploiting the embedded control circuit.

Step 1
Recognize on the laboratory LCD screen, the I/O pins and fill a table with a description for each signal.

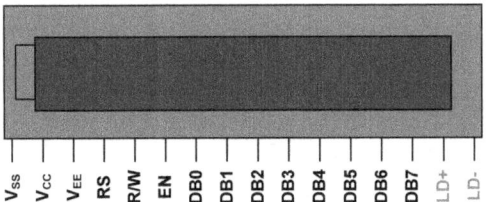

The available LCD screen of the laboratory may consist of 14 or 16 pins.

Step 2
Implement the following circuit connections and observe the contrast level by rotating the potentiometer.

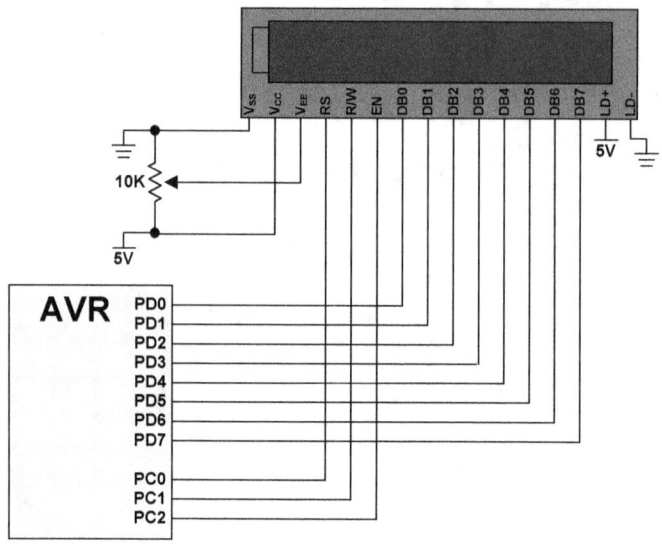

Step 3
Based on a chapter application, develop a program for displaying the message "Hello World" on the LCD screen.

Step 4
Use an oscilloscope to observe the change of the signal EN (Enable). Perform measurements and draw in a scaled paper the real waveform. Make comments for the real waveform as compared to the theoretical (as described in the corresponding application).

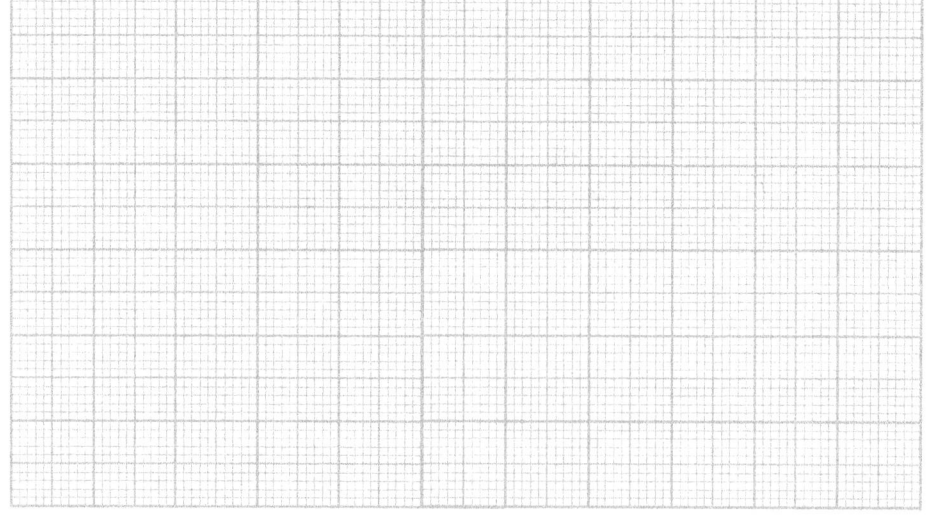

7 Switch circuits for user input

Content-Goals
Due to the fact that modern applications support input made by the users, it is necessary to build a multiple switch circuit for supporting input to the microcontroller. A multiple switch circuit can be used as a keyboard for entering numbers and symbols. In this chapter, the development of multiple switch circuits that can be used as a programmable keyboard for entering data will be presented and analyzed.

Chapter contents
7.1 Introduction
7.2 The keyboard layout

7.1 Introduction

The adaptation of applications to current user requirements and settings is a crucial feature that has to be efficiently supported. As mentioned in the previous chapter, the SSD units convert the microcontroller outputs to lighted digits in order to be readable by the users. On the other hand, multiple switch circuits are needed for supporting the user input process. A multiple switch circuit represents actually a button layout which has the role of a keyboard. In most of the cases, many buttons are needed for entering numbers and symbols (e.g. digits 0 to 9 or symbols 'A' to 'Z').

In practical level, this means that many pins are needed for controlling so many buttons. Moreover, an important goal is the minimization of the pins that will be used.

7.2 The keyboard layout

It is assumed that a 16 keys keyboard will be designed and final developed. The 16 keys are buttons that will be organized in a 4x4 layout (four lines and four columns). Using the proper code, buttons state will be always checked.

Figures 7.1a and 7.1b show two 4x4 keyboard layouts that can be found in the market.

234 ☐ CHAPTER 7

Figure 7.1a
4x4 keyboard layout

Figure 7.1b
4x4 keyboard layout

The buttons (switches) operation and circuit are based on the logic that have been presented in previous chapter. Figure 7.1c shows the known switch-button circuit. Using many times this circuit, a 4x4 layout can be developed (fig. 7.1d). Of course, there are many alternative implementations of the 4x4 layout, while a suitable program has to control the corresponding operation.

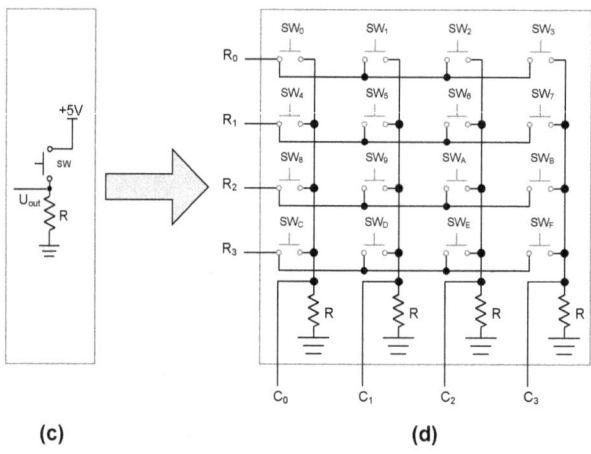

(c)   (d)

Figure 7.1 Switch and 4x4 layout circuit

If for example, a 5V signal is applied at R0, then the same voltage is applied in the whole line, due to the fact that the pins are horizontally connected. At the same time, this voltage will be also appeared at the pin C0 (if the button is pressed) which is vertically connected through the button. In other words, when a switch is closed (pressed button), then the the column where belongs is activated. If the activated line and column is known, then the pressed button is also known. Table 7.1, shows the corresponding active line and column for every individual switch-button.

Table 7.1 Line and column activation

Active line	Active column	Active switch
R_0	C_0	SW_0
R_0	C_1	SW_1
R_0	C_2	SW_2
R_0	C_3	SW_3
R_1	C_0	SW_4
R_1	C_1	SW_5

R_1	C_2	SW_6
R_1	C_3	SW_7
R_2	C_0	SW_8
R_2	C_1	SW_9
R_2	C_2	SW_A
R_2	C_3	SW_B
R_3	C_0	SW_C
R_3	C_1	SW_D
R_3	C_2	SW_E
R_3	C_3	SW_F

Figure 7.2 shows the «active-electric path» based on the SW_0 activation. Initially, the desired line is activated (e.g. R_0) and all the columns are scanned. Thus, by activating the line R_0 and button SW0, only the column C_0 gives 5V signal. With the line-column combination, the activated button is known. For finding any activated button, the lines R_0 to R_3 are connected to output pins of the microcontroller, while the columns C_0 to C_3 are connected to input pins of the microcontroller. In other words, when a line is activated through a microcontroller pin (output), then the column signals (column outputs) are read through the microcontroller pins (input).

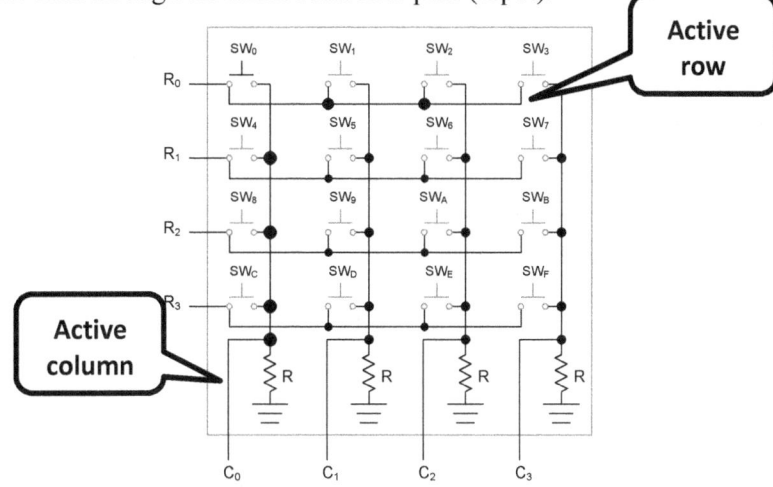

Figure 7.2 SW_0 is activated

The following algorithm describes the scanning operation in a circuit with N lines and M columns (NxM) for finding the pressed button.

```
BEGIN
For A=1 to N (lines)
  {
      Line A activation
      For B=1 to M (columns)
        {
            If column B=active then
                  the button (A,B) is pressed,
                  perform the corresponding operations
            End-If
        }
  }
END
```

As an example, the scanning operation of the columns C_0 to C_3, with the first line (R_0) activated will be analyzed.

Checking line R_0
    Set line R_0 to logic 1 (HIGH)

    Scan the logic level of C_0
    If $C_0=1$, then the SW_0 is pressed, exit

    Scan the logic level of C_1
    If $C_1=1$, then the SW_1 is pressed, exit

    Scan the logic level of C_2
    If $C_2=1$, then the SW_2 is pressed, exit

    Scan the logic level of C_3
    If $C_3=1$, then the SW_3 is pressed, exit
End of check

Application 7.1 – Developing a keyboard for displaying the digits 0 to 7 on SSD units

By knowing the switch-button circuit as well as the operation logic, the corresponding code for a 16 keys (4x4) keyboard will be developed. As shown in figure 7.3, the keys represent the symbols of the hexadecimal system (0 to F).

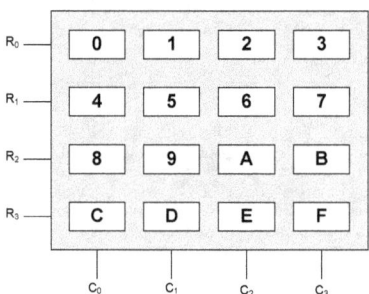

Figure 7.3 Hexadecimal keyboard

Behind the keyboard of figure 7.3 is the previous circuit (fig. 7.2), while figure 7.4 shows how the keyboard is connected with the microcontroller.

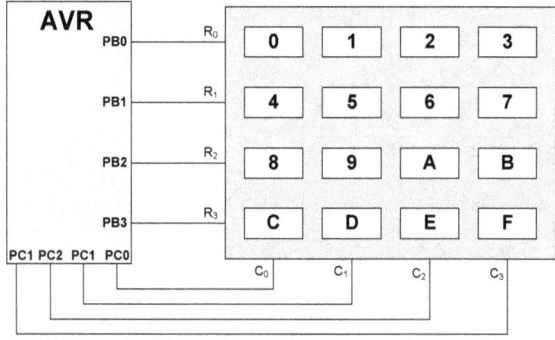

Figure 7.4 Keyboard and microcontroller interconnection

For verifying the keyboard operation as well as the corresponding code operation, the SSD units will be used for displaying the keyboard symbols. Figure 7.5a shows the full circuit which contains the input keyboard as well as the SSD units. Alternatively, only one SSD unit can be directly used (fig. 7.5b). The display circuit is based on the digital multiplexing of multiple SSD units. In this application, only the right side SSD unit (M0) is used for displaying the symbols. Thus the other SSD units (M1 to M3) will remain unconnected. This happens for showing the full circuit and the capability to display more digits later by using more SSD units.

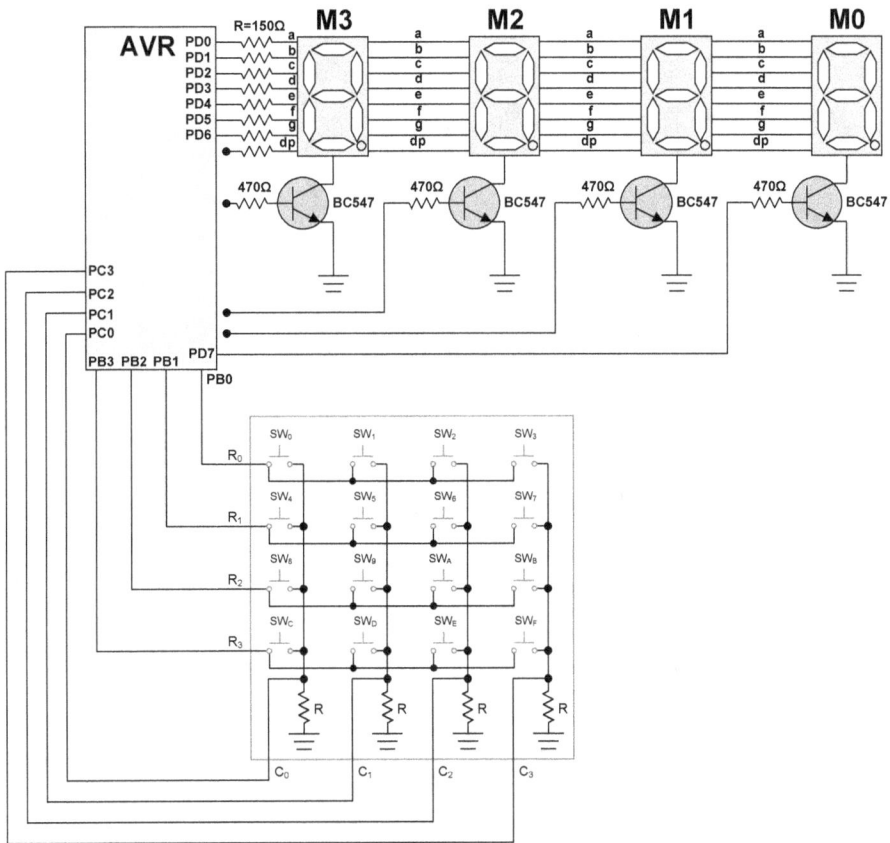

Figure 7.5a Test circuit for the 4x4 keyboard

As shown in the circuit of figure 7.5a, only the control line for the last digit is used (SSD unit M0), while the floating point led (dp) is also not used. Thus, the first seven pins of the port D are used for the segments a,b,c,d,e,f and g, while the last pin is used for controlling the SSD unit through the transistor. Due to the fact that the SSD unit is always active, the pin PD7 will be always 1 (fig. 7.6). The same connection method is also used in the circuit of figure 7.5b. Moreover, eight bits will be written to port D and thus the corresponding hexadecimal numbers for every digit will be different. Table 7.2 shows the hexadecimal numbers that correspond to the segments activation for displaying the digits 0 to 7 as well as the final number that will be used with MSB=1.

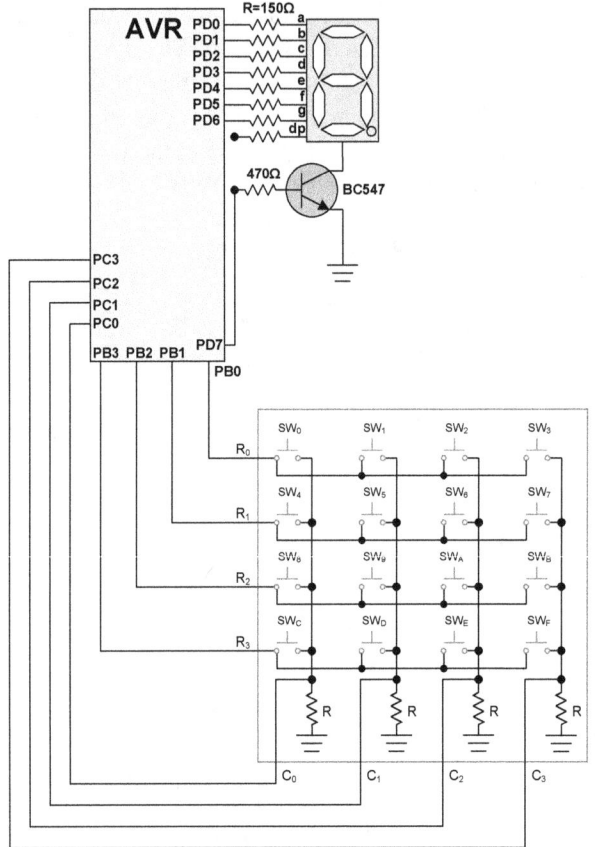

Figure 7.5b Test circuit with one SSD unit

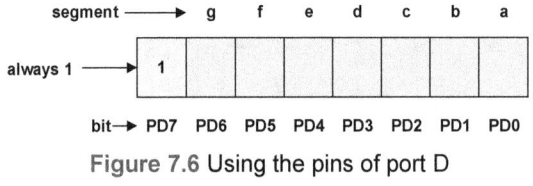

Figure 7.6 Using the pins of port D

Table 7.2 Digits display with MSB=1

Display digit	Binary (segments a to g)	Hexadecimal	Binary (with MSB=1) (segments a to g)	Hexadecimal
0	00111111	3F	10111111	BF
1	00000110	06	10000110	86
2	01011011	5B	11011011	DB
3	01001111	4F	11001111	CF
4	01100110	66	11100110	E6
5	01101101	6D	11101101	ED

| 6 | 01111101 | 7D | 11111101 | FD |
| 7 | 00000111 | 07 | 10000111 | 87 |

The first code (that will be developed) works for the two first lines of the keyboard. That means that only the numbers 0 to 7 (0-3 for the first line and 4-7 for the second line) will be displayed. As an example, figure 7.7 shows the "electrical path" when SW$_3$ is activated.

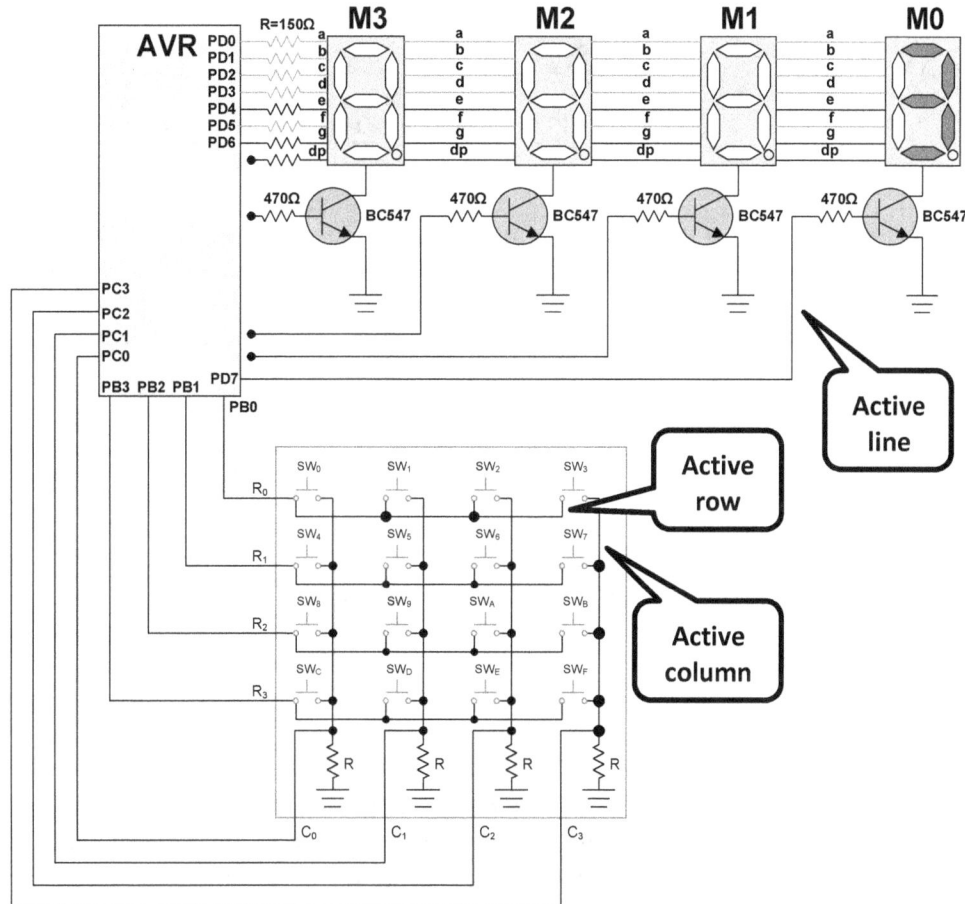

Figure 7.7 "Electrical path" for SW$_3$

As mentioned before, a test code will be developed for displaying the digits 0 to 7 by using the tow first lines of the keyboard. The first code that will be developed is very simple and contains operations that are repeated within separated code segments. This is done in order to understand the basic operation principle and to make later improvements regarding the code size and the corresponding operation. The first code is analyzed in steps regarding the needed operations.

Step 1. Port initialization

All the pins of the port D are initialized as outputs (0xFF=0b11111111, fig. 7.8).

```
LDI R16,0xFF
OUT DDRD,R16
```

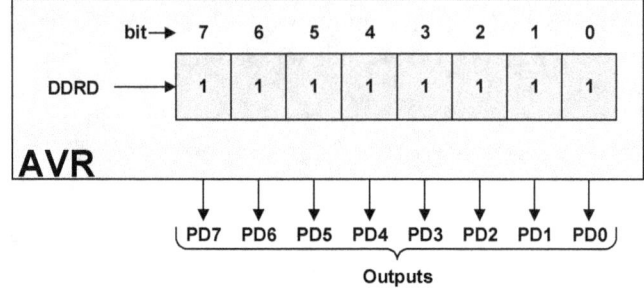

Figure 7.8 Port D settings

```
OUT DDRB,R16
```

The same value (0xFF) is loaded in register DDRB, for initializing the corresponding pins as outputs (fig. 7.9). This is done for simplicity reasons despite that only the pins PB0 and PB1 are used (line 0 and 1 respectively).

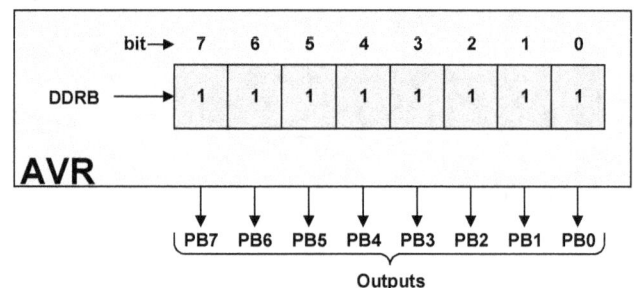

Figure 7.9 Port B settings

```
CBI DDRC,0
CBI DDRC,1
CBI DDRC,2
CBI DDRC,3
```

Now, the pins PC0 to PC3 are initialized as inputs (despite that the initial content of the register DDRC is zero). In this section, the modification method for specific pins is demonstrated by loading the zero value (fig. 7.10). The same could be done also for the port B.

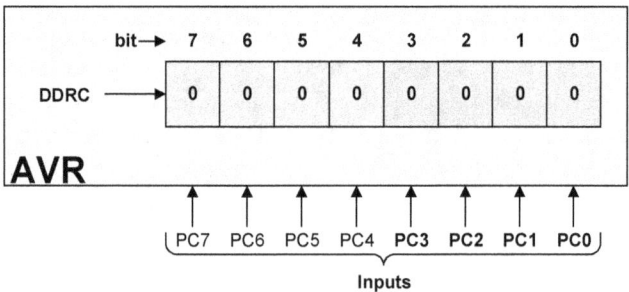

Figure 7.10 Port C settings

Switch circuits for user input ☐ 241

Step 2. Activating the first line (R0)

```
LDI   R16,0b00000001
OUT   PORTB,R16
nop
```

All the pins of port B are initialized for giving the capability to access the whole register PORTB which sets the voltage level in the corresponding outputs. Knowing that the least significant bit corresponds to the pin PB0, the value 0b00000001 will be loaded in the register in order to activate the R_0. After the PB0 activation, an 1μSec delay is implemented (for 1MHz clock) before the columns state reading. This is done in order to be enough time for the external circuit to set the signal level before the columns reading.

Figure 7.11 shows the activation of the first line (R_0) through the output pin PB0.

Step 3. Reading the columns state

```
IN R17,PINC
```

Next, the pin status that are connected with the columns is read (actually, all the pins are scanned even those that are not used).

Step 4. Checking the fourth column

```
MOV   R16,R17
ANDI  R16,0b00001000
CPI   R16,0b00001000
BREQ  key3
```

When a line is activated and a button is pressed, then the corresponding column is also activated. For finding the pressed button, the active line and column who gives logic 1 (5V) have to be known. This signal comes from the pin of port B which is connected to the active line, while the logic 1 is transferred to the pin which is connected to the corresponding column through the external circuit.

Figure 7.12 shows the "electrical path" which is activated from the line R_0 and the column C_3 through the SW_3.

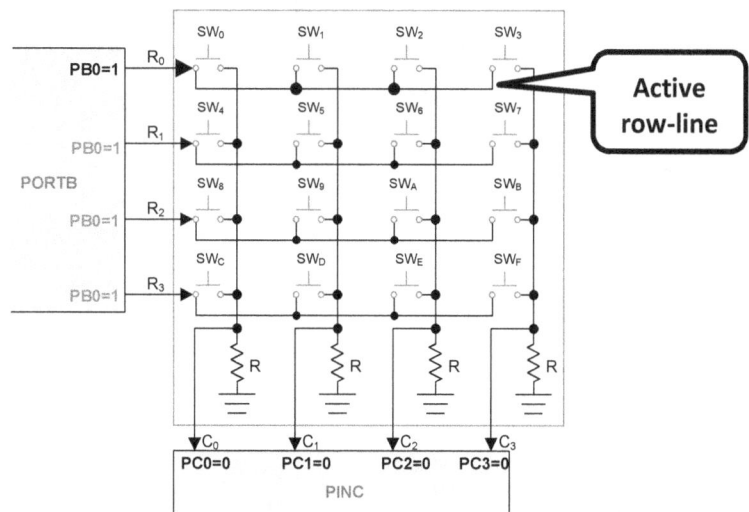

Figure 7.11 Activating the first line

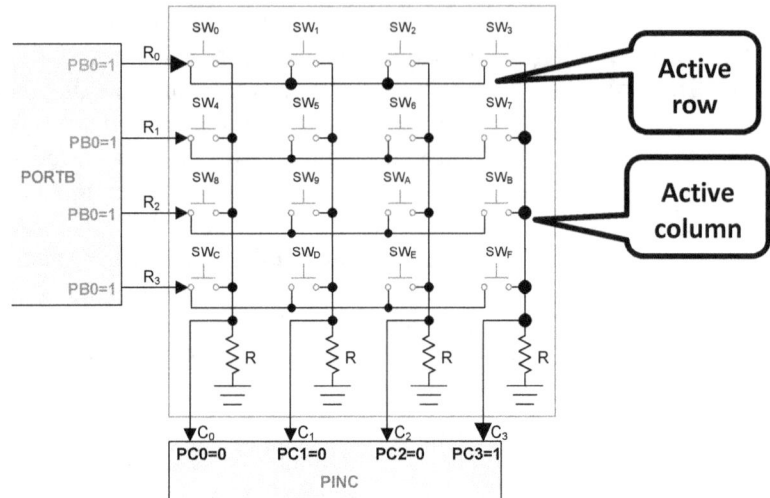

Figure 7.12 Detecting the SW₃

Next, with the instruction `MOV R16,R17` the content of PINC (port C) is temporarily transferred for processing. This is done because the R17 has to be used with its initial value for other checking.

A method to check if the last column is activated, is to use the mask 00001000 with a logical AND to the bits that are read from the port. A logical AND with zero produces always a zero as a result.

Thus, the logical AND is focused on the specific bit, while the other bits are cleared. If the logical AND is performed with 1, then if the specific bit is zero, then remains zero, but if is 1, then remains 1.

In other words, if the specific bit is activated, then the logical operation gives the same state as a result. This equivalence is checked with the instruction `CPI` (comparing a register with an integer value). If this equivalence is TRUE, then the condition of the instruction `BREQ` is TRUE and the execution flow is continued from the label `key3`.

Step 5. Displaying the digit

```
key3:
LDI R19,3
RCALL print_digit
RJMP return


print_digit:
   LDI   ZL,LOW(digits*2)
   LDI   ZH,HIGH(digits*2)

   ADD   ZL,R19
   LPM

   OUT   PORTD,R0
   RET
```

As fifth step, the execution flow from the label `key3` will be analyzed (as a successive step from the previous check) in order to show what is happening when the button is pressed. The code which starts from the label `key3` calls the subroutine `print_digit` in order to display the corresponding digit on the SSD unit.
The value of the digit is defined through the register `R19`. Actually, the content of the register `R19` is used as a distance pointer from the beginning of the array `digits` which contains the bit values for activating the correct segments of the SSD unit.

```
digits: .DB 0xBF, 0x86, 0xDB, 0xCF, 0xE6, 0xED, 0xFD, 0x87
```

The complete testing code follows.

Code 7.1

```
;Place here the proper INC file for your microcontroller model
;(if needed), e.g. ATmega32/ATmega32A => m32def.inc/m32Adef.inc,
;ATmega328 => m328def.inc
.INCLUDE "include/m32Adef.inc"

;***************************
;Stack initialization
;***************************
LDI R16,HIGH(RAMEND)
OUT SPH,R16
LDI R16,LOW(RAMEND)
OUT SPL,R16


;***************************
;Port initialization
;***************************
                        ;Ports for the SSD units
LDI R16,0xFF            ;R16=11111111
OUT DDRD,R16            ;All the pins of port D, outputs

                        ;Ports for the lines
OUT DDRB,R16            ;All the pins of port B, outputs

                        ;Ports for the columns
CBI DDRC,0              ;Pin PC0, input
CBI DDRC,1              ;Pin PC1, input
CBI DDRC,2              ;Pin PC2, input
CBI DDRC,3              ;Pin PC3, input
```

```
;*******************************
;button scanning (keyboard)
;*******************************
start:
                            ;Activate first line
LDI  R16,0b00000001         ;Load the proper bits for the pins
OUT  PORTB,R16              ;Write to port for activation
NOP                         ;Delay
IN   R17,PINC               ;Read columns status

                            ;Check for column 3
MOV  R16,R17                ;Transfer the Column status
                            ;(temporarily)
ANDI R16,0b00001000         ;Logical AND with the mask
CPI  R16,0b00001000         ;Check to find
                            ;if the SW3 is pressed
BREQ key3                   ;Change execution flow to
                            ;label key3 (true case)


                            ;Check for column 2
MOV  R16,R17                ;Transfer the Column status
                            ;(temporarily)
ANDI R16,0b00000100         ;Logical AND with the mask
CPI  R16,0b00000100         ;Check to find
                            ;if the SW2 is pressed
BREQ key2                   ;Change execution flow to
                            ;label key2 (true case)


                            ;Check for column 1
MOV  R16,R17                ;Transfer the Column status
                            ;(temporarily)
ANDI R16,0b00000010         ;Logical AND with the mask
CPI  R16,0b00000010         ;Check to find
                            ;if the SW1 is pressed
BREQ key1                   ;Change execution flow to
                            ;label key1 (true case)


                            ;Check for column 0
MOV  R16,R17                ;Transfer the Column status
                            ;(temporarily)
ANDI R16,0b00000001         ;Logical AND with the mask
CPI  R16,0b00000001         ;Check to find
                            ;if the SW0 is pressed
BREQ key0                   ;Change execution flow to
                            ;label key0 (true case)


                            ;Activate second line
LDI  R16,0b00000010         ;Load the proper bits for the pins
OUT  PORTB,R16              ;Write to port for activation
NOP                         ;Delay
```

```
            IN  R17,PINC               ;Read columns status

                                       ;Check for column 3
            MOV  R16,R17               ;Transfer the Column status
                                       ;(temporarily)
            ANDI R16,0b00001000        ;Logical AND with the mask
            CPI  R16,0b00001000        ;Check to find
                                       ;if the SW7 is pressed
            BREQ key7                  ;Change execution flow to
                                       ;label key7 (true case)

                                       ;Check for column 2
            MOV  R16,R17               ;Transfer the Column status
                                       ;(temporarily)
            ANDI R16,0b00000100        ;Logical AND with the mask
            CPI  R16,0b00000100        ;Check to find
                                       ;if the SW6 is pressed
            BREQ key6                  ;Change execution flow to
                                       ;label key6 (true case)

                                       ;Check for column 1
            MOV  R16,R17               ;Transfer the Column status
                                       ;(temporarily)
            ANDI R16,0b00000010        ;Logical AND with the mask
            CPI  R16,0b00000010        ;Check to find
                                       ;if the SW5 is pressed
            BREQ key5                  ;Change execution flow to
                                       ;label key5 (true case)

                                       ;Check for column 0
            MOV  R16,R17               ;Transfer the Column status
                                       ;(temporarily)
            ANDI R16,0b00000001        ;Logical AND with the mask
            CPI  R16,0b00000001        ;Check to find
                                       ;if the SW4 is pressed
            BREQ key4                  ;Change execution flow to
                                       ;label key4 (true case)

            RJMP start                 ;Return to the start

                                       ;Display the digit 7
                                       ;on the SSD unit
            key7:
            LDI R19,7
            RCALL print_digit
            RJMP return
                                       ;Display the digit 6
                                       ;on the SSD unit
            key6:
            LDI R19,6
            RCALL print_digit
```

```
RJMP return
                        ;Display the digit 5
                        ;on the SSD unit
key5:
LDI R19,5
RCALL print_digit
RJMP return
                        ;Display the digit 4
                        ;on the SSD unit
key4:
LDI R19,4
RCALL print_digit
RJMP return
                        ;Display the digit 3
                        ;on the SSD unit
key3:
LDI R19,3
RCALL print_digit
RJMP return
                        ;Display the digit 2
                        ;on the SSD unit
key2:
LDI R19,2
RCALL print_digit
RJMP return

                        ;Display the digit 1
                        ;on the SSD unit
key1:
LDI R19,1
RCALL print_digit
RJMP return
                        ;Display the digit 0
                        ;on the SSD unit

key0:
LDI R19,0
RCALL print_digit
                        ;Reference point for
                        ;return to the start
return:
RJMP start

;*******************************
;Display digit subroutine
;*******************************
print_digit:
  LDI   ZL,LOW(digits*2)
  LDI   ZH,HIGH(digits*2)
  ADD   ZL,R19
  LPM
```

```
        OUT    PORTD,R0
        RET

;Digits 0 to 7
digits: .DB 0xBF, 0x86, 0xDB, 0xCF, 0xE6, 0xED, 0xFD, 0x87
```

Application 7.2 – Displaying the digits 0 to F on SSD units using the keyboard

In this application, the 4x4 keyboard will be used for displaying the hexadecimal digits on the SSD units. By studying the previous application it is obvious that many code segment operations are repeated. That means that the same process is implemented for finding the activation of every button. As will be shown, the code of this application is much shorter as compared to the previous one despite the fact that exploits all the buttons of the keyboard. Figure 7.13 shows the flow chart diagram for the code that will be developed.

The code development based on the flow chart diagram is as follows:

(1)
```
LDI R16,0xFF
OUT DDRD,R16
```
By writing the number 0xFF (11111111 in binary), all the pins of port D are set as outputs. Through these outputs the digit display on the SSD units is controlled.

```
OUT DDRB,R16
```
The same operation is applied also in port B (all the pins are outputs). Through port B, the desired line is activated.

248 ☐ CHAPTER 7

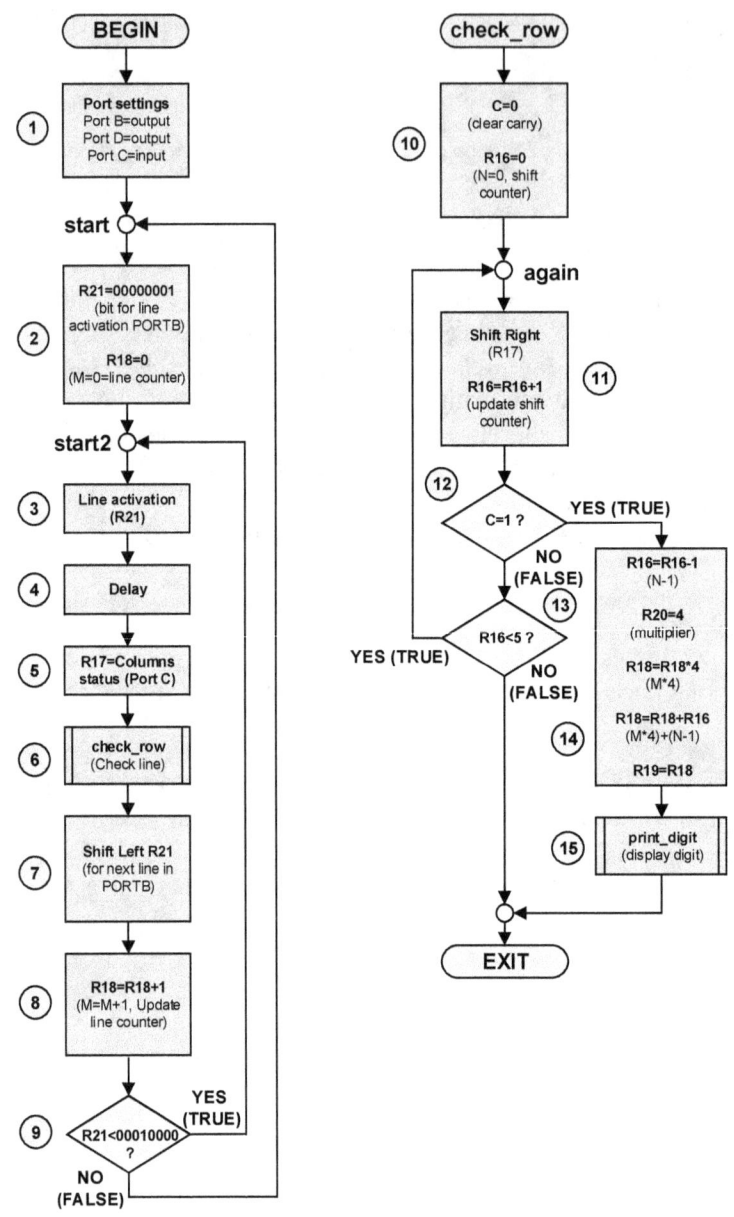

Figure 7.13 Flow chart diagram

```
CBI DDRC,0
CBI DDRC,1
CBI DDRC,2
CBI DDRC,3
```

For studying once again the setting of specific pins, the instruction CBI is used for clearing the corresponding bits of the register DDRC. With this code segment, the pins 0 to 3 of the port C are set as inputs. Through port C, the activated column is detected.

(2)

```
start:
LDI R21,0b00000001
LDI R18,0
```
The process begins from the first line. The register `R21` holds the needed bit value which will be written to `PORTB` in order to activate the line. For the first line, the value 00000001 will be used. The bit value from the right first bit to the fourth on the left corresponds to the lines (rows) 0 to 3.

The register `R18` will be used as counter of the current line in order to calculate the digit that must be displayed on the SSD unit based on the pressed button.

(3)
```
start2:
OUT PORTB,R21
```
Starting the main process, the line is activated by loading the content of `R21` in the register `PORTB`.

(4)
```
nop
```
Time delay for signal stabilization in the external circuit before the columns read.

(5)
```
IN R17,PINC
```
Reading the columns state through the register `PINC` (port C).

(6)
```
RCALL check_row
```
Calling the subroutine for columns scanning (regarding the current line)

(7)
```
LSL R21
```
Shifting left of the register `R21` content. The register `R21` holds an '1' which corresponds to the active line. For example, the value 00000001 means that the line 0 (first line) will be activated. Thus, for activating the next line the value 00000010 has to be loaded in the register, etc. The decimal numbers that correspond to the values that have to be loaded for successively activating the keyboard lines are shown in table 7.3.

Table 7.3 PORTB content

Binary number (8bit)	Activated line	Hexadecimal
00000001	0 (first)	1
00000010	1 (second)	2
00000100	2 (third)	4
00001000	3 (fourth)	8

As shown above, an arithmetic calculation has to be performed in order to load successively the numbers 1,2,4 and 8 in the register R21. Actually, is about a multiplication with 2. This can be easily done by one left shift (instruction `LSL`). For example, if the 00000001 has been initially loaded, then with a left shift, the '1' will be moved one position to the left while the least significant bit will be filled with a zero. Thus, the number 00000001 (value 1) with one shift will be 00000010 (value 2), in order to activate the next line.

(8)
```
INC R18
```
After the update of register R21 for activating the next line, the current line counter is also updated (register R18).

(9)
```
CPI R21,0b00010000
BRLT start2
```
The keyboard consists of four lines. The first line is activated by loading the number 00000001 (in the register R21) and the fourth by loading the number 00001000. After the process execution for the third line, a left shift takes place for one more time. Thus, the new content of the register R21 will be 00010000. Due to the fact that there is no fifth line, the process will be terminated and thus the execution flow will return to the label start2 (activating the next line) if the content of the register R21 is less than the value 00010000 (values 00000001, 00000010, 00000100 and 00001000).

(10), (11)
```
check_row

CLC
LDI R16,0
again:
      LSR R17
      INC R16
```

The content of register PINC which represents the state of columns has been loaded in the register R27. The final code that has been implemented, works automatically for every individual line. Based on the content of register PINC, the activated line is known. The content of register PINC based on the activated column is shown in the table 7.4.

Table 7.4 PINC content

Binary number (8bit)	Activated column	Hexadecimal
00000001	0 (first)	1
00000010	1 (second)	2
00000100	2 (third)	4
00001000	3 (fourth)	8

For finding automatically the activated column, a shift right is performed using the carry. In this type of shift, the left part is filled with zeros, while the rejected bit is stored in the carry. As an example, it is assumed that the third column is activated. In such a case, the content of PINC will be 00000100. After the loop completion, the register R16 contains the number of shifts that have been performed. Figure 7.14

shows the successive shifts. The shifts are stopped when the carry becomes 1. Thus, the number of activated column is equal to the number of shifts.

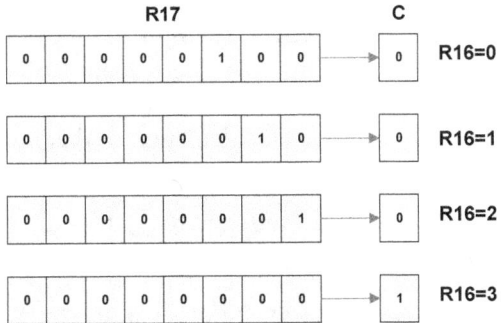

Figure 7.14 Shifts for calculating the activated column

As shown in figure 7.14, the carry becomes 1 (C=1) after 3 shifts. In other words, the number of shifts depends on the position of the '1'. This position corresponds to the activated column. Thus, if 3 shifts have been performed, then the 3rd column has been activated. The shift operation is performed with the instruction LSR, while after every shift the counter value is increased (register R16).

(12)
```
BRCS exit
```
If the carry is activated, then the process is terminated because the activated column has been found. The execution flow is continued from the label exit for displaying the proper digit on the SSD unit (this is performed with the instruction BRCS).

(13)
```
CPI R16,5
    BRLT again
```
Due to the fact that the code has to work for any case, special care has to be taken if no column is activated. This means that the content of the register R17 is zero. If R17 is zero, then the check condition for the carry status is not enough and as a result, the loop is never terminated. For this reason, a second condition is placed in order not to allow more than 4 shifts (the columns are only 4). Thus, even the carry is not '1' (R17=0), the process is terminated after 4 iterations. If the loop exit is performed through this condition, then no column has been activated and the execution flow returns (RET) from the subroutine check_row.

(14)
```
exit:

DEC R16
LDI R20,4
MUL R18,R20
MOV R18,R0
ADD R18,R16
MOV R19,R18
```

At this point, a last step has to be done before the final digit display on the SSD unit. As mentioned before, the needed process for scanning the keyboard and displaying the corresponding number is repeated automatically for every new active line. Reaching this code section (label `exit`, exit from the previous loop due to carry activation) the active line (line counter, `R16`) and column (column counter, `R18`) are already known.

The important issue here is the automatic calculation of the digit which corresponds to the active line and column. The digit will be calculated from the following formula:

$$D=(M*4)+(N-1)$$
Where **D**=digit, **M**=active line, **N**=active column

For the above formula it is known that M=R18 and N=R16.

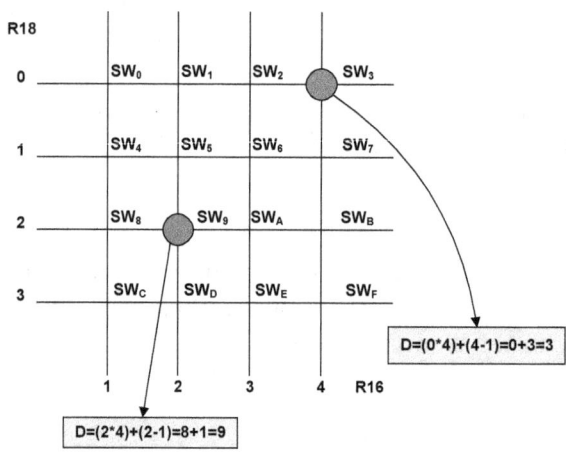

Figure 7.15 Digit calculation

Figure 7.15 shows two examples of calculating the content of the registers R16 and R18. If for example the button SW$_9$ is pressed, then R18=2, R16=2 and D=9. In other words, the digit 9 has to be displayed due to the fact that the button 9 is pressed. The same calculation is performed also for the button 3. It must be noticed also the usage of the register R0 after the multiplication. In the case of the instruction MUL, the low byte of the result is stored automatically in the register R0.

(15)
`RCALL print_digit`
At this point the process for the pressed button is completed, due to the fact that the subroutine for the digit display (`print_digit`) is called. As mentioned before, the digit is used as a distance pointer within the `digits` array for displaying the correct number on the SSD unit.

The completed code follows.
Code 7.2

```
;Place here the proper INC file for your microcontroller model
;(if needed), e.g. ATmega32/ATmega32A => m32def.inc/m32Adef.inc,
;ATmega328 => m328def.inc

.INCLUDE "include/m32Adef.inc"

;***************************
;Stack initialization
;***************************
LDI R16,HIGH(RAMEND)
OUT SPH,R16
LDI R16,LOW(RAMEND)
OUT SPL,R16

;******************************
;Ports initialization
;******************************
LDI R16,0xFF
OUT DDRD,R16         ;Port for the SSD unit

OUT DDRB,R16         ;Port for line activation

                     ;Port for columns checking
CBI DDRC,0
CBI DDRC,1
CBI DDRC,2
CBI DDRC,3


;******************************
;Main program
;******************************
start:
LDI R21,0b00000001   ;Load bit value for the first line
LDI R18,0            ;Initialize line counter
start2:

OUT PORTB,R21        ;Set the port B pins, as outputs
NOP                  ;Delay (1µSec at 1MHz)
IN R17,PINC          ;Read the columns status

RCALL check_row      ;Checking the current line

LSL R21              ;Left shif for the next line
INC R18              ;Increase line counter
CPI R21,0b00010000   ;Check if the third line is exceeded
BRLT start2          ;Return to start2 if all the
                     ;lines have not been activated

RJMP start           ;Return to the start for
                     ;scanning all the lines
```

```
;*******************************
;Checking line subroutine
;*******************************
check_row:
CLC                         ;Clear carry
LDI R16,0                   ;Initialize shift counter
again:
      LSR R17               ;Shift right by one position
      INC R16               ;Increase shift counter
      BRCS exit             ;If the '1' is "arrived" to carry,
                            ;then calculate the digit for display
      CPI R16,5             ;Check for the maximum number of shifts
      BRLT again            ;If all the shifts are not completed,
                            ;then return to again for
                            ;the next shift
RET                         ;Return of the carry is not 1
exit:                       ;Process the activated column
                            ;N=R16, row=R18
                            ;Calculate D
                            ;D=row*4+(N-1)
  DEC R16                   ;calculate N-1
  LDI R20,4                 ;Load multiplier (R20=4)
  MUL R18,R20               ;R0=row*4
                            ;(if the result can be
                            ;stored in 1 byte)
  MOV R18,R0                ;Read the result from the R0
  ADD R18,R16               ;Final calculation of D
                            ;D=R18=(row*4)+(N-1)
  MOV R19,R18               ;Load the result to R19 for
                            ;calling the display digit
                            ;subroutine
  RCALL print_digit         ;Calling the subroutine
  RET                       ;Return (complete the process)

;*******************************
;Display digit subroutine
;*******************************
print_digit:
  LDI    ZL,LOW(digits*2)
  LDI    ZH,HIGH(digits*2)
  ADD    ZL,R19
  LPM
  OUT    PORTD,R0
  RET

digits: .DB 0xBF, 0x86, 0xDB, 0xCF, 0xE6, 0xED, 0xFD, 0x87, 0xFF,
0xEF, 0xF7, 0xFC, 0xD8, 0xDE, 0xF9, 0xF1
```

Figures 7.16 to 7.31 show the symbols 0 to F on the SSD unit.

Figure 7.16

Figure 7.17

Figure 7.18

Switch circuits for user input 255

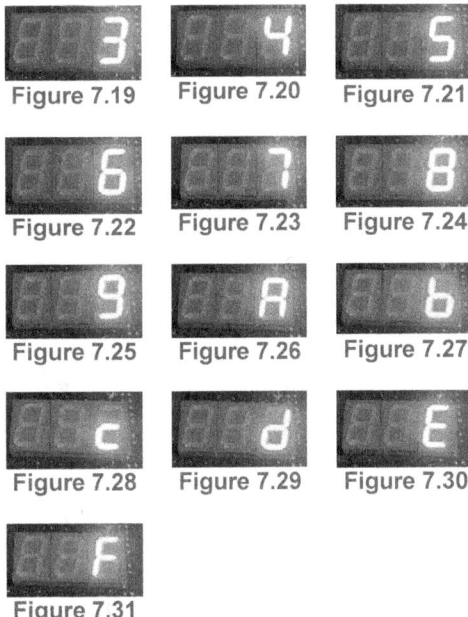

Figure 7.19 Figure 7.20 Figure 7.21

Figure 7.22 Figure 7.23 Figure 7.24

Figure 7.25 Figure 7.26 Figure 7.27

Figure 7.28 Figure 7.29 Figure 7.30

Figure 7.31

LABORATORY EXERCISE 6
Keyboard development

GOAL
To understand the method of developing switch matrices (organized button group) which constitutes a keyboard and supports user input.

Step 1
Write the signals and the necessary read method for determining the activation of the switch-button SW_5.

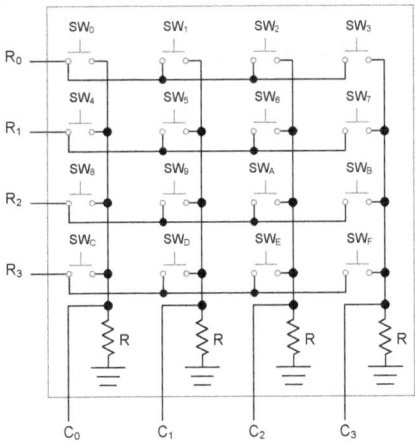

Step 2
Implement the corresponding circuit based on the following circuit diagram.

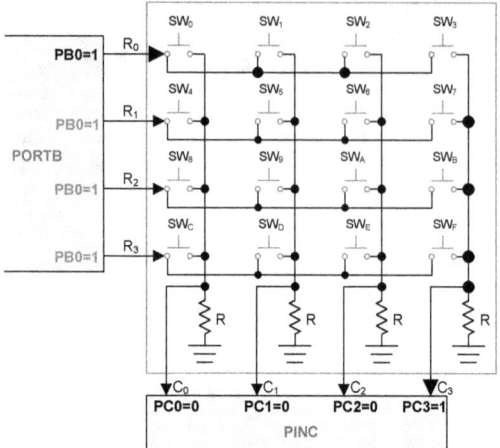

Step 3
Modify the corresponding application code in order to activate the LEDs LED0 to LED3 when the buttons 0, 5, A, or F are pressed.

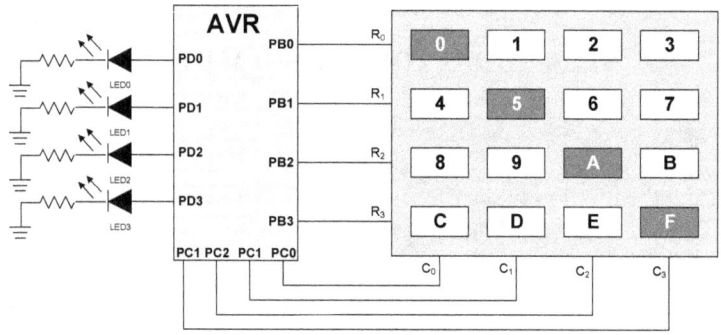

Step 4 (homework)
Develop an application for supporting the input of a four-digit number. Every time, when a digit is entered, the displayed digits on the SSD units will be shifted by one position to the left (until the four digits to be filled).

A. Simulating Assembly source code in Atmel Studio 7

Introduction

Atmel (now Microchip) offers the software development environment Atmel Studio (AS) for free. Within the AS, the application source code (e.g. Assembly, C) can be developed and simulated. During the simulation process, the content of registers or memory, ports status, etc., can be viewed. Moreover, break points can be placed in any selected instructions for supporting the execution step by step. The simulation process is quite accurate due to the fact that is based on the real microcontroller datasheet. The microcontroller model is selected during the project creation process.

Creating and simulating a microcontroller project

The process for creating and simulating a microcontroller project is as follows:

STEP 1 Start Atmel Studio

Atmel Studio 7.0

Figure A.1 Software shortcut

The Atmel Studio software can be executed through the applications menu or from the corresponding shortcut within the Desktop location (fig, A.1). After the **double click on the shortcut**, the following splash screen is appeared (fig. A.2).

Figure A.2 Splash screen

**STEP 2
Start a new project**

Figure A.3 shows the startup page of the AS. The user has the main option to start a new project or to open a recently saved project. Also, selected updates can be installed from the notification area. The notifications are clickable options for direct update using the internet browser or by selected specific extension to install.

For creating a new project, click the option **New Project...** from the Start area (on the left, fig. A.3).

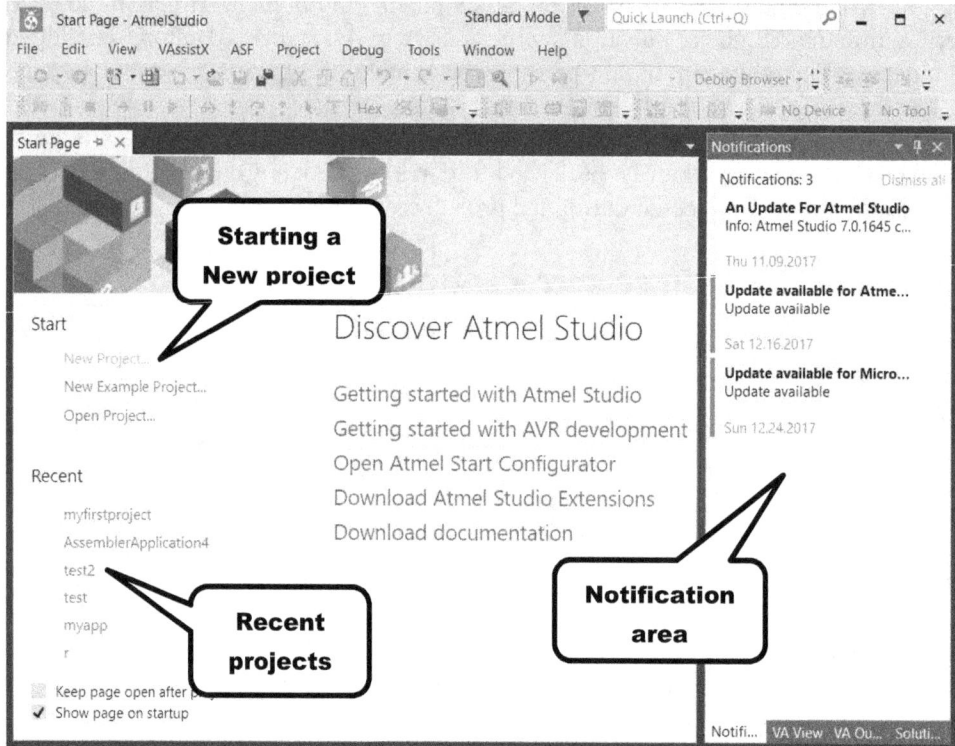

Figure A.3 Startup page

**STEP 3
Create project**

As already known the applications of the book are based on assembly language. Thus, the new project that will be created will also be based on assembly language. In the 'New Project' window, enter the following parameters (fig. A.4):
(a) Project type: Select **AVR Assembler Project**
(b) Project name: Set a **project name** of your choice (e.g. testproject2)

(c) Location: Set the **working folder** (be sure that you have permissions to write in that folder)

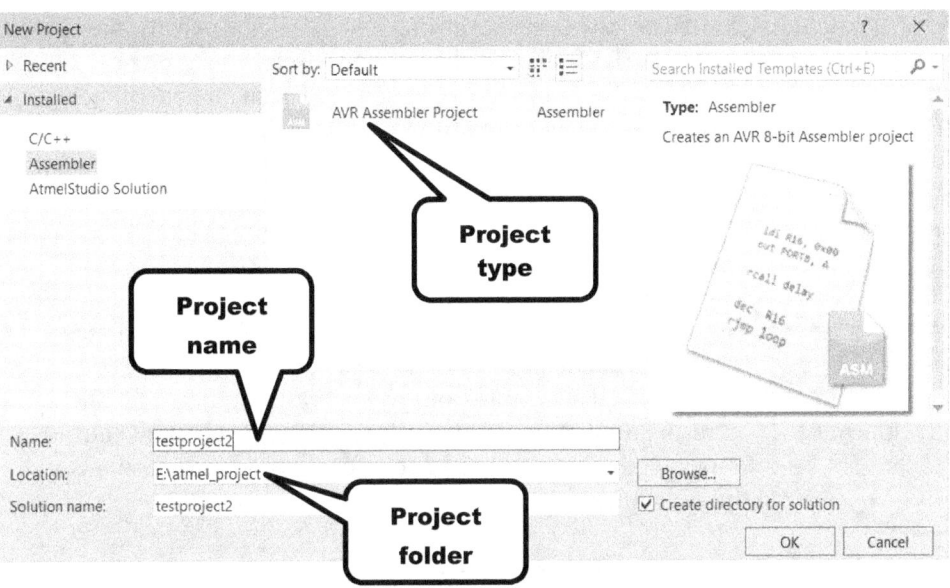

Figure A.4 Creating a project

STEP 4 Select AVR model

As mentioned in previous chapter, an INC file has to be included in the source code in order to use easily the capabilities of the selected AVR model. In Atmel Studio it is not necessary because this is done by selecting the AVR model (fig. A.5) during the project creation process.

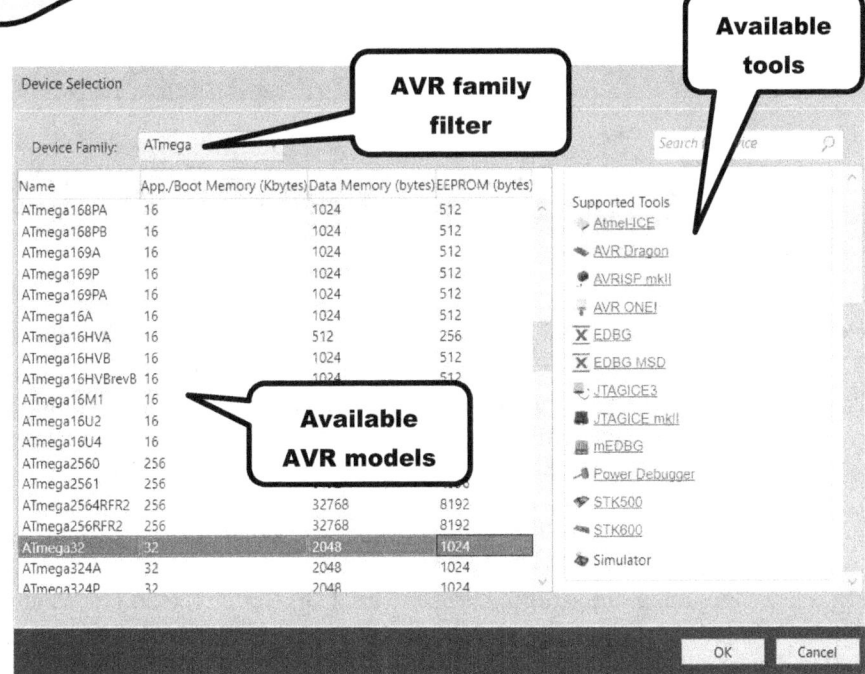

Figure A.5 Selecting the AVR model

As shown in figure A.5 a filter can be used in order to find more easily the desired microcontroller model. On the right side of the "Device Selection", there are some available external tools which can be connected to computer. The main goal of these tools is to host a microcontroller model (physical chip) which can be programmed within the Atmel Studio. It must be noticed that only certain hardware tools are supported by the Atmel Studio. In the same window, the basic features such as the memory size (e.g. program/data memory) of the selected model are shown. From the available list, **select Atmega32** for developing the corresponding application.

STEP 5 Code development

The source code of the application is entered within the folder **main.asm** (fig. A.6). The following sample code (fig. A.6) sets all the pins of port B as outputs and the corresponding pins in the state 10101010. **Enter your source code** (or the sample code of figure A.6) and **press the save button**.

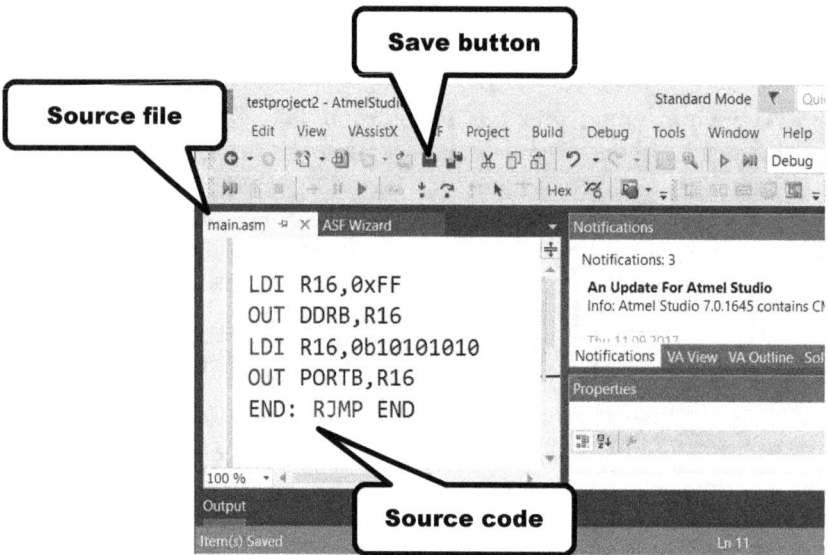

Figure A.6 Entering source code

STEP 6 Build solution

The next step is to build solution. By this process, the HEX (executable for the target microcontroller) will be generated. During this process, the source code is checked for syntactic and other errors (e.g. right instruction usage-syntax). For starting the building process, **select Build Solution** (or press F7) from the **menu Build**.

Information messages about the building process are displayed in the output window (fig. A.7). Figure A.7 also shows that the building process regarding the sample source code was completed successfully.

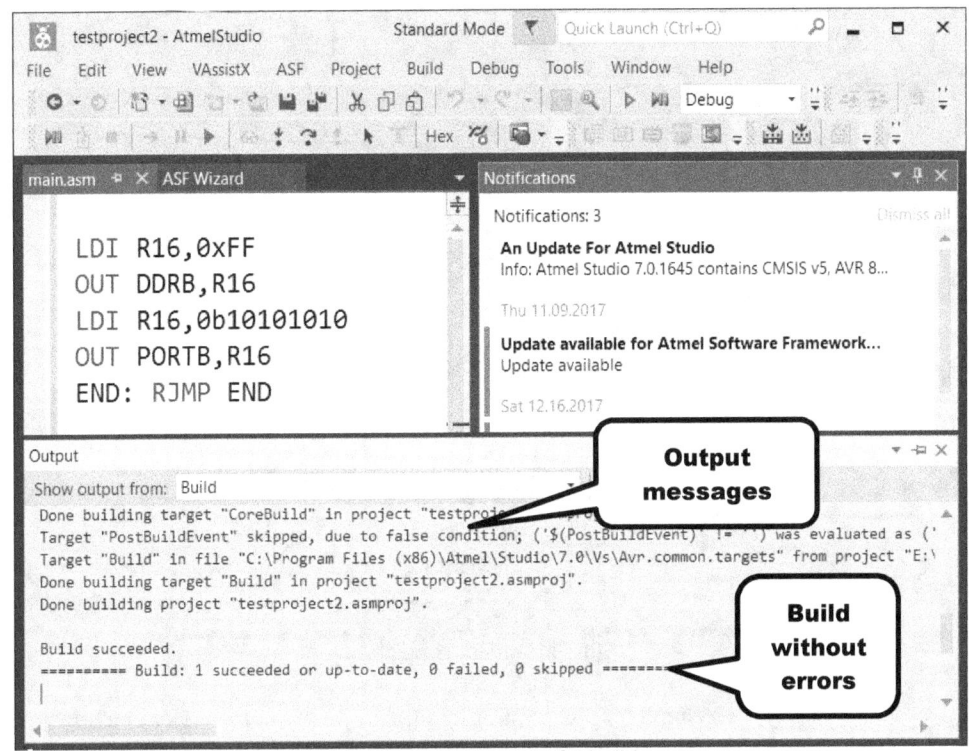

Figure A.7 Successful build

The simulation environment of AS, offers the capability to watch the registers content, the port status, the status register content, etc., during the code execution.

For studding the impact of every assembly instruction to an element of the microcontroller, break points are used. If a break point is assigned to an instruction, then the execution flow is stopped for watching internal information of the microcontroller (e.g. registers content). The execution flow is continued by the user action. Thus, a step by step execution can be performed for giving the time to user for studying in depth the corresponding code operation. **Make a left click on the left side of each instruction.** If the break point is placed successfully, then a small red cycle is displayed on the left side of instruction.

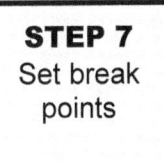

STEP 7
Set break points

Figure A.8 shows the placed break points in the sample source code.

It must be noticed that after every modification, *the Save button has to be pressed. If the symbol '*' is displayed on the folder title of active window, then a modification has been detected from the AS and thus, a Save operation has to be performed.*

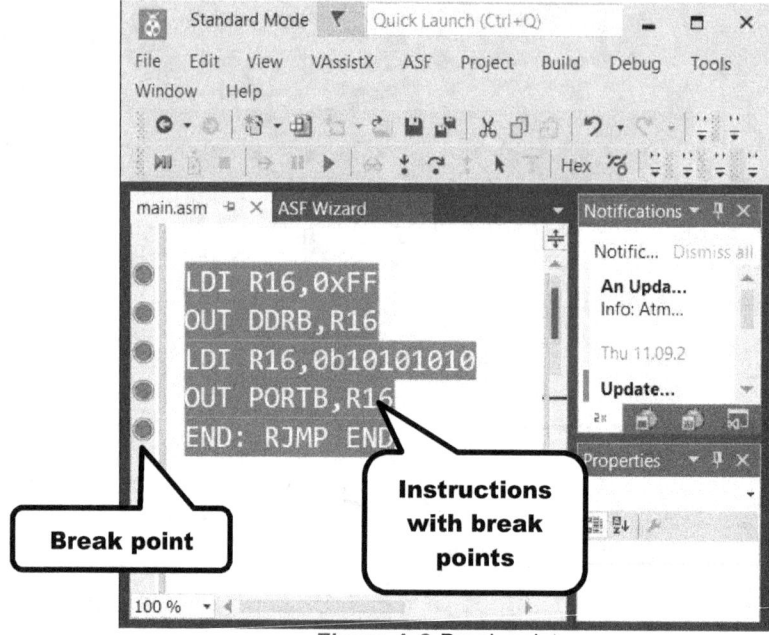

Figure A.8 Break points

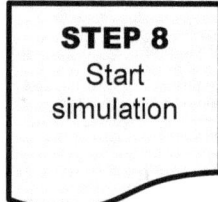

STEP 8
Start simulation

The next step is to simulate the code execution. For starting this operation, the button play has to be pressed (fig. A.9).

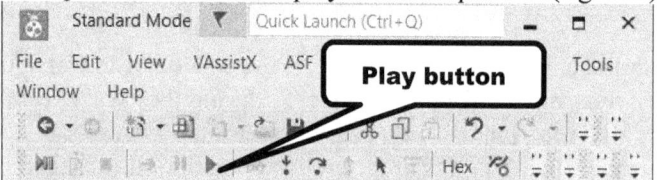

Figure A.9 Start simulation

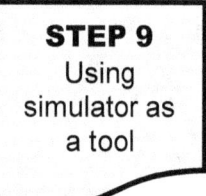

STEP 9
Using simulator as a tool

If no physical board is available (fig. A.10), the program execution can be studied by the embedded simulation tool.

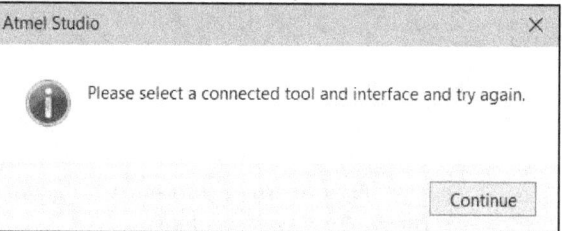

Figure A.10 No tool is selected

After the play button is pressed, AS asks the user to select a tool for continuing the process. **Select Simulator** (fig. A11) and press the **Save button**.

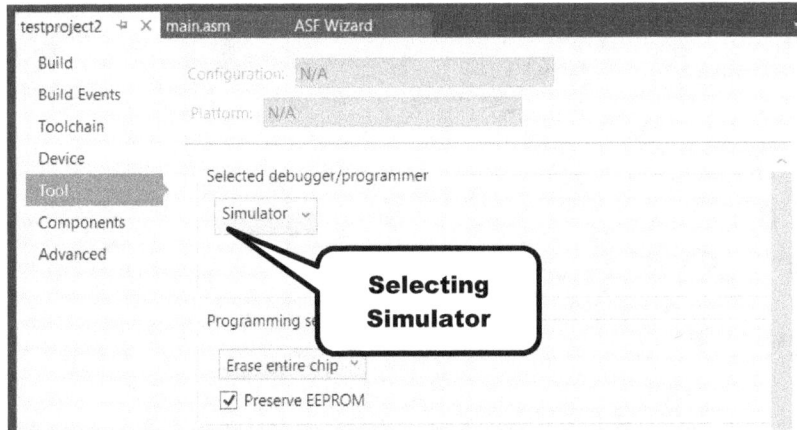

Figure A.11 Selecting the simulator tool

STEP 10
Run the simulation step by step

Now the execution process can be started. **Press successively the play button** in order to view the corresponding results in the microcontroller. Figure A.12 shows the information which is displayed during the step by step execution. From the menu **Debug**, select **Windows** to activate the information window type.

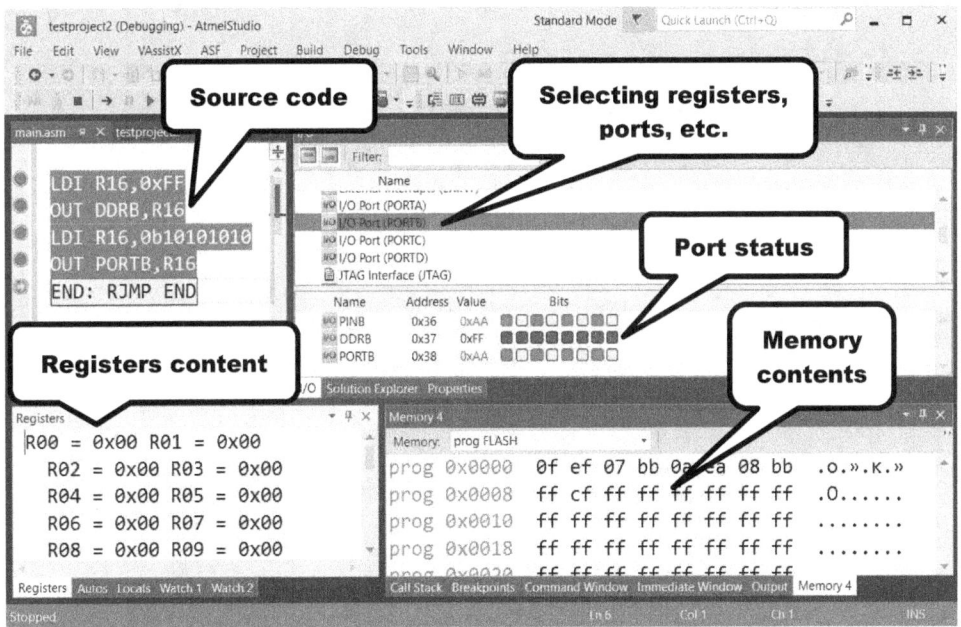

Figure A.12 Simulation process

B Instruction Summary

Load instructions

Instruction	Description	Operation	Bits of the SREG that are affected
MOV Rd,Rs	Loading from register to register (8bit) d,s ∈ [0,31]	Rd=Rs	-
MOVW Rd,Rs	Loading from register to register (16bit) d,s ∈ {0,2,...,30}	Rd+1:Rd=Rs+1:Rs	-
LDI Rd,k	Loading an integer to a register (integer value) d ∈ [16,31], k ∈ [0,255]	Rd=k	-
LD Rd,X	Loading (indirect) from a memory location (pointed by X) to a register d ∈ [0,31]	Rd=(X)	-
LD Rd,X+	Loading (indirect) from a memory location (pointed by X) to a register and X increment by 1 (after) d ∈ [0,31]	Rd=(X) X=X+1	-
LD Rd,-X	X decrement by 1 (before) and Loading (indirect) from a memory location (pointed by X) to a register d ∈ [0,31]	X=X-1 Rd=(X)	-
LD Rd,Y	Loading (indirect) from a memory location (pointed by Y) to a register d ∈ [0,31]	Rd=(Y)	-
LD Rd,Y+	Loading (indirect) from a memory location (pointed by Y) to a register and Y increment by 1 (after) d ∈ [0,31]	Rd=(Y) Y=Y+1	-
LD Rd,-Y	Y decrement by 1 (before) and Loading (indirect) from a memory location (pointed by Y) to a register d ∈ [0,31]	Y=Y-1 Rd=(Y)	-
LDD Rd,Y+q	Loading (indirect) from a memory location (pointed by Y+q) to a register (q is an integer value) d ∈ [0,31]	Rd=(Y+q)	-
LD Rd,Z	Loading (indirect)	Rd=(Z)	-

Instruction	Description	Operation	Flags
	from a memory location (pointed by Z) to a register $d \in [0,31]$		
LD Rd,Z+	Loading (indirect) from a memory location (pointed by Z) to a register and Z increment by 1 (after) $d \in [0,31]$	Rd=(Z) Z=Z+1	-
LD Rd,-Z	Z decrement by 1 (before) and Loading (indirect) from a memory location (pointed by Z) to a register $d \in [0,31]$	Z=Z-1 Rd=(Z)	-
LDD Rd,Z+q	Loading (indirect) from a memory location (pointed by Z+q) to a register (q is an integer value) $d \in [0,31]$	Rd=(Z+q)	-
LDS Rd,addr	Loading (direct) a register from a memory location (RAM) $d \in [0,31]$	Rd=(addr)	-
ST X,Rs	Loading (indirect) from a register to a memory location (pointed by X) $s \in [0,31]$	(X)=Rs	-
ST X+,Rs	Loading (indirect) from a register to a memory location (pointed by X) and X increment by 1 (after) $s \in [0,31]$	(X)=Rs X=X+1	-
ST -X,Rs	X decrement by 1 (before) and Loading (indirect) a register to a memory location (pointed by X) $s \in [0,31]$	X=X-1 (X)=Rs	-
ST Y,Rs	Loading (indirect) from a register to a memory location (pointed by Y) $s \in [0,31]$	(Y)=Rs	-
ST Y+,Rs	Loading (indirect) from a register to a memory location (pointed by Y) and Y increment by 1 (after) $s \in [0,31]$	(Y)=Rs Y=Y+1	-
ST -Y,Rs	Y decrement by 1 (before) and Loading (indirect) a register to a memory location (pointed by Y) $s \in [0,31]$	Y=Y-1 (Y)=Rs	-
STD Y+q,Rs	Loading (indirect) from a register to a memory location (pointed by Y+q), q is an integer value $s \in [0,31]$	(Y+q)=Rs	-
ST Z,Rs	Loading (indirect) from a register to a memory location (pointed by Z) $s \in [0,31]$	(Z)=Rs	-
ST Z+,Rs	Loading (indirect)	(Z)=Rs	-

Instruction	Description	Operation	
	from a register to a memory location (pointed by Z) and Z increment by 1 (after) s ∈ [0,31]	Z=Z+1	
ST −Z,Rs	Z decrement by 1 (before) and Loading (indirect) a register to a memory location (pointed by Z) s ∈ [0,31]	Z=Z−1 (Z)=Rs	−
STD Z+q,Rs	Loading (indirect) from a register to a memory location (pointed by Z+q), q is an integer value s ∈ [0,31]	(Y+q)=Rs	−
STS addr,Rs	Loading (direct) a memory location (RAM) from a register s ∈ [0,31]	(addr)=Rs	−

Arithmetic and Logical instructions

Instruction	Description	Operation	Bits of the SREG that are affected
ADD Rd,Rs	Add two registers (without carry) d,s ∈ [0,31]	Rd=Rd+Rs	H,S,V,N,Z,C
ADC Rd,Rs	Add two registers with carry d,s ∈ [0,31]	Rd=Rd+Rs+C	H,S,V,N,Z,C
ADIW Rd,k	Add an integer value to a 16bit registers pair d ∈ {24,26,28,30}, k ∈ [0,63]	Rd+1:Rd=Rd+1:Rd+k	S,V,N,Z,C
SUB Rd,Rs	Subtract two registers (without carry) d,s ∈ [0,31]	Rd=Rd−Rs	H,S,V,N,Z,C
SUBI Rd,k	Subtract an integer value from a register d ∈ [16,31], k ∈ [0,255]	Rd=Rd−k	H,S,V,N,Z,C
SBC Rd,Rs	Subtract two registers with carry d,s ∈ [0,31]	Rd=Rd+Rs−C	H,S,V,N,Z,C
SBCI Rd,k	Subtract an integer value and the carry from a register d ∈ [16,31], k ∈ [0,255]	Rd=Rd−k−C	H,S,V,N,Z,C
SBIW Rd,k	Subtract an integer value from a 16bit registers pair d ∈ {24,26,28,30}, k ∈ [0,63]	Rd+1:Rd=Rd+1:Rd−k	S,V,N,Z,C
AND Rd,Rs	Logical AND between two registers d,s ∈ [0,31]	Rd=Rd AND Rs	S,V=0,N,Z
ANDI Rd,k	Logical AND between a register and an integer value d ∈ [16,31], k ∈ [0,255]	Rd=Rd AND k	S,V=0,N,Z
OR Rd,Rs	Logical OR between registers d,s ∈ [0,31]	Rd=Rd OR Rs	S,V=0,N,Z
ORI Rd,k	Logical OR between a register and an integer value d ∈ [16,31], k ∈ [0,255]	Rd=Rd OR k	S,V=0,N,Z

Instruction	Description	Operation	Bits of the SREG that are affected
EOR Rd,Rs	Logical EXCLUSIVE OR between two registers d,s ∈ [0,31]	Rd=Rd XOR Rs	S,V=0,N,Z
COM Rd	Logical NOT (one's complement) of a register content d ∈ [0,31]	Rd=0xFF-Rd	S,V=0,N,Z,C=1
NEG Rd	Two's complement calculation of a register content d ∈ [0,31]	Rd=0x00-Rd	H,S,V,N,Z,C
INC Rd	Increment the register content by 1 d ∈ [0,31]	Rd=Rd+1	S,V,N,Z
DEC Rd	Decrement the register content by 1 d ∈ [0,31]	Rd=Rd-1	S,V,N,Z
CLR Rd	Clear the register content d ∈ [0,31]	Rd=0x00	S=0,V=0,N=0,Z=1
SER	Load the maximum value on a register d ∈ [16,31]	Rd=0xFF	-
MUL Rd,Rs	Unsigned numbers multiplication through registers d,s ∈ [0,31]	R1:R0=Rd*Rs	Z,C
MULS Rd,Rs	Signed numbers multiplication through registers d,s ∈ [16,31]	R1:R0=Rd*Rs	Z,C
MULSU Rd,Rs	Signed/Unsigned numbers multiplication trhough registers d,s ∈ [16,23]	R1:R0=Rd*Rs	Z,C
FMUL	Unsigned fractioned numbers multiplication d,s ∈ [16,23]	R1:R0=Rd*Rs	Z,C
FMULS	Signed fractioned numbers multiplication d,s ∈ [16,23]	R1:R0=Rd*Rs	Z,C
FMULSU	Multiplication of a signed fractioned number with an unsigned fractioned number d,s ∈ [16,23]	R1:R0=Rd*Rs	Z,C

Shift and bit manipulation instructions

Instruction	Description	Operation	Bits of the SREG that are affected
LSL Rd	Logical shift left d ∈ [0,31]	C=Rd(7) Rd(n+1)=Rd(n) Rd(0)=0	Z, C, N, V, H
LSR Rd	Logical shift right d ∈ [0,31]	C=Rd(0) Rd(n)=Rd(n+1) Rd(7)=0	Z, C, N, V
ROL Rd	Shift left through carry d ∈ [0,31]	Rd(0)=C Rd(n+1)=Rd(n) C=Rd(7)	Z, C, N, V, H

ROR Rd	Shift right through carry d ∈ [0,31]	Rd(7)=C Rd(n)=Rd(n+1) C=Rd(0)	Z, C, N, V
ASR Rd	Arithmetical shift right through carry d ∈ [0,31]	Rd(n)=Rd(n+1), n=0..6	Z, C, N, V
SWAP Rd	Swap High/Low Parts of a register d ∈ [0,31]	Rd(3..0)⇔Rd(7..4)	-
BSET s	Sets a specific bit in the SREG s ∈ [0,7]	SREG(s) = 1	SREG(s)
BCLR s	Clears a specific bit in the SREG s ∈ [0,7]	SREG(s) = 0	SREG(s)
SEC	Sets the bit C of SREG	C=1	C
CLC	Clears the bit C of SREG	C=0	C
SEN	Sets the bit N of SREG	N=1	N
CLN	Clears the bit N of SREG	N=0	N
SEZ	Sets the bit Z of SREG	Z=1	Z
CLZ	Clears the bit Z of SREG	Z=0	Z
SEI	Sets the bit I of SREG	I=1	I
CLI	Clears the bit I of SREG	I=0	I
SES	Sets the bit S of SREG	S=1	S
CLS	Clears the bit S of SREG	S=0	S
SEV	Sets the bit V of SREG	V=1	V
CLV	Clears the bit V of SREG	V=0	V
SET	Sets the bit T of SREG	T=1	T
CLT	Clears the bit T of SREG	T=0	T
SEH	Sets the bit H of SREG	H=1	H
CLH	Clears the bit H of SREG	H=0	H

www.ingramcontent.com/pod-product-compliance
Lightning Source LLC
Chambersburg PA
CBHW062212220526
45471CB00009B/3170